TURKEY IN THE MIDDLE EAST

Published under the auspices of the
Harry S. Truman Research Institute for the Advancement of Peace,
The Hebrew University of Jerusalem

TURKEY IN
THE MIDDLE EAST

OIL, ISLAM, AND POLITICS

ALON LIEL

TRANSLATED BY EMANUEL LOTTEM

LYNNE
RIENNER
PUBLISHERS

BOULDER
LONDON

Published in the United States of America in 2001 by
Lynne Rienner Publishers, Inc.
1800 30th Street, Boulder, Colorado 80301
www.rienner.com

and in the United Kingdom by
Lynne Rienner Publishers, Inc.
3 Henrietta Street, Covent Garden, London WC2E 8LU

Library of Congress Cataloging-in-Publication Data
Liel, Alon.
   [Turkiyah ba-Mizrah ha-Tikhon. English]
   Turkey in the Middle East : oil, Islam, and politics / Alon Liel ; translated by
Emanuel Lottem.
     p. cm.
   Includes bibliographical references and index.
   ISBN 1-55587-909-8
   1. Turkey—Foreign economic relations—Middle East. 2. Middle East—Foreign
economic relations—Turkey. 3. Petroleum industry and trade—Government
policy—Turkey. 4. Turkey—Foreign relations—Middle East. 5. Middle East—Foreign
relations—Turkey. 6. Islam and politics—Turkey. I. Title.
HF1583.4.Z4 M628513 2001
327.561056—dc21   2001019006

British Cataloguing in Publication Data
A Cataloguing in Publication record for this book
is available from the British Library.

Printed and bound in the United States of America

5   4   3   2   1

Contents

Preface   vii

1   Introduction: Between East and West   1

Part 1   Turkey's Dependence on Oil

2   The Economics of Energy in Turkey   27

3   The Diplomacy of Cheap Oil   53

4   Bartering for Oil   79

5   International Consequences of the 1980 Crisis   103

6   The 1980s: Happy Days Are Here Again   117

**Part 2   Turkey in the Middle East:
International Aspects**

7   The Muslim Option   129

8   Politics, Oil, and Islam:
    Relations with Iraq, Iran, and Libya   155

9   Turkey and the Arab-Israeli Conflict   189

Part 3   Turkey Toward the Twenty-first Century

10   The Islamic Challenge   219

11   Groundswell, Landslide, and Earthquake   233

List of Acronyms   245
Selected Bibliography   247
Index   249
About the Book   253

Preface

This book is the product of my more than twenty years of research on Turkish and regional affairs, both as a member of the Israeli diplomatic service and as an academic scholar.

During the three decades with which the book is mainly concerned, 1970–2000, an extraordinary transformation took place in Turkey's Middle Eastern policies. For example, Turkey's relations with Israel, nearly meaningless in the 1970s, became intensely cooperative in the 1990s. This transformation reflected the renewed Western orientation of Turkey's foreign policy, which in turn was mainly the result of the country's recovery from the severe economic crisis of the late 1970s, following the global energy crisis. This story will be told here in some detail.

I acknowledge with thanks the support provided by the Rothschild and Yad HaNadiv Foundations, the Dayan Center at Tel Aviv University, and the Department of International Relations at the Hebrew University in Jerusalem, my academic home these past few years. Special thanks to Amnon Cohen, Victor Azarya, and Edy Kaufman, members of the Truman Institute, whose help made possible the publication of the English-language version of the book.

I also express my thanks to those Turkish diplomats and politicians who consented to be interviewed for this study, including former Turkish prime minister and president Süleyman Demirel. Many thanks to members of Israel's diplomatic staff in Turkey, including Ambassadors Zvi Alpeleg and Uri Barner. To Turkey's previous ambassador to Israel, Barlas Özener, I am grateful for many conversations into the wee hours on Turkey and its marvels.

Finally, I wish to thank HaKibbutz HaMeuhad Publishing House, the translator and editor, Emanuel Lottem, and Lynne Rienner, who initiated publication in the United States.

Before this version of this book went to print, I lost my (and the translator's) friend Aryeh Dvir, a member of the Israeli Foreign Ministry and a hero of the Six Day War, who made many helpful comments on the first version of the book. May he rest in peace.

—*Alon Liel*

I

INTRODUCTION:
BETWEEN EAST AND WEST

The Anatolia plateau, the heartland of the Republic of Turkey, has for millennia been home to numerous tribes, peoples, and empires—a variegated patchwork of cultures and religions. Anatolia's geopolitical position between East and West has played a major role in shaping its inhabitants' systems of government and views of the world around them.

The Kemalist revolution that so radically transformed Turkey in the 1920s from an Islamic empire into a secular republic led Turkey toward the modern West culturally, economically, and politically. But as Turkey was becoming a secular state, its population never shrugged off its religious beliefs and heritage. After the death of revolutionary leader Mustafa Kemal Atatürk in 1938, the Turks showed a marked determination to preserve their Islamic identity and develop good relations with neighboring Muslim nations while retaining the heritage of their revered leader. This inclination fit well, particularly during the 1970s, with Turkey's economic exigencies and its desire not to become overly dependent on the West.

Turkey had become interested in having good relations with its Arab neighbors in the 1920s and 1930s, at a time when most Arab countries were still under foreign rule; since then, Turks have followed closely those countries' struggle for independence. First contacts were made in the 1930s with Egypt and particularly with Iraq, a highly important neighbor. Republican Turkey took pains to disavow any territorial claims, especially to the Mosul area, which became part of Iraq after World War I. Turkey and Iraq even joined the Sa'adabad Treaty in July 1937, alongside Iran and Afghanistan. That same year Turkey launched Radio Ankara's Arabic-language broadcasts to inform the Arab world of its views. Turkey's leader, İsmet İnönü, used the inaugural occasion to send a message of friendship to the Arab nations.

Still, Turkey's relations with the Arab world could not be easily cleared of residue from the Ottoman period, most notoriously during its last few years. The Arabs could not forget the harsh oppression of the nascent Arab national movements, and the Turks did not readily forget how the Arabs had betrayed them during World War I. The Kemalist reforms of the 1920s and 1930s made the establishment of normal relations with the Arab countries all the more difficult since their underlying themes, Westernization and modernization, were anathema to the Arabs.

Immediately after World War II, when most Arab countries gained independence, conditions became more favorable for a Turkish-Arab rapprochement, but only superficially so. In effect, no close relationships developed except with Iraq, and that for only a short time. The period 1945–1965 was marked by tensions in Turkey's relations with most Arab countries, notably Egypt and Syria. One of the reasons for such tension was Turkey's wholehearted association with the West, especially the United States, at the time. In 1947 Turkey supported the Truman Doctrine; in July 1948 it was included in the Marshall Plan (to the tune of $3 billion); it fought with the West in the Korean War; and, most ominously, it joined the North Atlantic Treaty Organization (NATO) in February 1952, allowing a massive presence of U.S. troops on its territory.

The chief bone of contention between Turkey and most Arab states was its support during the 1950s of U.S. efforts to create a series of regional defense treaties aimed against the Soviet Union. A few Arab capitals believed the Baghdad Treaty—formed in 1955 by Turkey, Iraq, Iran, and Pakistan—served as a vehicle for a Turkish regional leadership position, a feeling that exacerbated existing tensions. In the Arab world in the 1950s a nationalist bloc was created, led by Egypt and Syria, that tended to support the Soviet position in the international arena and to oppose Turkey's regional ambitions. At the same time, since 1949 Turkey had been developing relations with Israel, which also raised little enthusiasm among the Arabs.

The year 1958 marked an ebb in Turkey's relations with the Arab nations. It was a stormy year in the Middle East: an anti-West coup took place in Iraq, Nasserism was strengthening with the establishment of the United Arab Republic, and civil war broke out in Lebanon. Turkey's decision to allow U.S. troop movements through its territory on their way to Lebanon as part of Western intervention in the civil war there strengthened Ankara's image as the West's only reliable ally in the Muslim Middle East. At the same time, however, Arabs regarded it as an anti-Arab move. All in all, except for Ankara's relations with Baghdad in the period 1955–1958, there were ongoing tensions in Turkey's relations with the Arab world during the 1950s.

The chill continued into the 1960s, but then a transformation took place. Particularly important among the reasons for this change were the general relaxation of interbloc tensions and the Turkish-Greek conflict in Cyprus.

As the Cold War began to thaw with the coming of détente, some Middle Eastern nations, having become less dependent on the superpowers, were able to orient their foreign policies in ways not readily available previously. Inside Turkey, voices were calling for greater freedom of action in foreign policy, especially toward the Third World and neighboring Arab states. From then on, Ankara's policies were governed by new, more nationalistic precepts, and relations with the West were no longer allowed to interfere with the development of relations with other Muslim nations and the USSR.

The Cyprus crisis broke out in 1964 as a result of growing violence between Greeks and Turks on the island and rapidly escalated into direct Turkish military intervention. Turkey's posture in the crisis caused tensions in relations with Washington. In June 1964 President Lyndon B. Johnson sent a letter to Prime Minister İnönü to inform him that should Turkish military involvement in Cyprus bring about Soviet intervention, NATO members would not regard themselves as committed to Turkey's defense. Turkish public opinion reacted strongly to the U.S. position; both government and the public began to reexamine Turkey's international position. Faced with increasing international preoccupation with the Cyprus crisis and frequent UN discussions, Turkey felt an urgent need for support, which it hoped to find in the Arab countries, the Eastern bloc, and the Third World. When Turkey finally moved into Cyprus militarily in July 1974, occupying one-third of the island's territory, U.S.-Turkish relations reached rock bottom. The United States imposed an arms embargo on Turkey, claiming U.S.-supplied weapons were used in operations in Cyprus in contravention of the terms under which Turkey had received U.S. military assistance in the first place. In response, the Turkish government suspended its defense treaty with the United States and closed down U.S. military installations on its territory. At the same time, indirect military assistance provided by Libya and Iraq during the Cyprus war (particularly jet fuel) made these nations extremely popular with the Turkish public.

First indications of closer relations with the Arab world in the mid-1960s included visits to Turkey by Tunisian president Habib Bourgiba (March 1965), Saudi King Feisal (August 1965), and Iraqi president 'Abd-al-Salaam 'Arif (February 1967). The pace of rapprochement was increased after the Six Day War when Jordan's King Hussein came to Ankara (September 1967), Turkey's prime minister Süleyman Demirel went to Iraq (November 1967), Turkey's president Cevdet Sunay visited Saudi Arabia, Libya, and Iraq (January–April 1968), and Morocco's King Hassan II visited Turkey (April 1968). Such high-level mutual visits among Jordan, Syria, Tunisia, Algeria, and Kuwait continued during the early 1970s. Further strengthening of Turkey's ties with the Muslim world came in September 1969 when the Turks accepted King Hassan's invitation to participate in the Islamic summit convened in protest of an act of arson at the al-Aqsa Mosque in Jerusalem.

Still, many in Turkey's establishment wished to be attached in the long run to the West rather than the Third World or the Muslim nations. Several Arab countries, particularly Saudi Arabia, saw the new Turkish policy of strengthening its ties with the Muslim world as political and economic opportunism. Thus Turkey's improved relations with the Arab world in the late 1960s and early 1970s cannot be regarded on the whole as a fundamental change in Ankara's Middle Eastern policy.

Under the special circumstances created by the energy crisis during the 1970s, however, Turkey drew nearer to the Arab world. Among Muslim nations, Turkey was hit hardest by the crisis. On several occasions between 1977 and 1981, its economy came close to total collapse because of severe and enduring shortages of fuel, power, and essential goods. Energy shortages caused grave damage to industry and agriculture, reduced transport to a near standstill, and paralyzed the education and health systems for long periods. The costs of such system failures to the economy as a whole were far greater than the actual cost of the fuel needed to sustain these systems. As an indication of how severely the economy was hurt, in 1980 crude petroleum import costs were 30 percent higher than all of Turkish export revenue, a condition without parallel among Western oil consumers.

This rapid increase in Turkey's oil import bills made it advisable to rely on more than one oil supplier. Besides Iraq, Turkey's traditional supplier, lobbying efforts often amounting to plain begging were directed at Saudi Arabia, Iran, and Libya during the period 1974–1980. Eventually, Turkey came to rely on the goodwill of three major oil suppliers: Iraq, Libya, and Iran.

Turkey's position following the 1973–1974 oil embargo was unusual. As a Muslim nation and a close neighbor of several oil producers, it could have based its oil imports on direct deals with those nations (unlike most other consumers, who had to rely on multinational corporations and the free market). On the other hand, as part of the Western world Turkey could have joined the efforts launched by oil consumers to curb Organization of Petroleum Exporting Countries (OPEC) price extravagances and prevent future oil shortages.

Walking this diplomatic tightrope, Turkey tried to enjoy the best of both worlds: its direct supply contracts with producing nations on the one hand and reliance on Western contingency plans on the other. Having created (at least on paper) this dual defense, however, Turkey found itself confronted with a severe and prolonged shortage of fuel, which brought about a lasting change not only in its Middle Eastern policies but also, to an extent, in its social and cultural fabric.

During the period 1974–1981, Turkey's ties with the Arab world strengthened to a remarkable extent. Following the 1974 oil embargo, its economic relations with the oil countries became the foundation of its Middle Eastern policy, whereas economic needs in general took the lead in policy decisionmaking as a whole. A secure oil supply, a rapid increase in export revenues to finance an ever growing oil bill, and employment overseas for surplus labor were vital concerns. This massive dependence on Muslim oil completely changed the patterns of Turkish foreign trade during the late 1970s and early 1980s. The Muslim world replaced the West as Turkey's chief export destination.

In the mid-1970s Turkey's leaders attempted to keep separate the country's growing need for oil and its foreign policy. But the inability to pay the ever increasing price demanded for oil made them eventually submit to a series of harsh demands by Iraq and Libya and later by Iran and Saudi Arabia. Initially, these demands were mainly economic, but later political and military demands were made as well. One result was a marked change in Turkey's Middle Eastern policy orientation. The once warm relationship between Turkey and Israel (as allies with the West) grew colder, becoming nearly meaningless by the late 1970s and remaining so during most of the 1980s. At the same time, as we shall see, the influence of Turkey's strengthening ties with the Muslim world extended from economics and politics into the very fabric of Turkey's society and culture. The main beneficiaries—both economically and politically—were Islamic groups in Turkey, which had long waited for such an opportunity.

An intriguing development in Turkey's relationship with the Muslim oil countries during the 1980s was a complete transformation in the balance of power that had prevailed during the 1970s. At that time, when Turkey was completely dependent on Iraq, Libya, and Iran (accumulating an oil debt of around $2 billion), issues such as debt rescheduling were among Turkey's most important policy considerations. During the 1980s, however, when oil prices were sagging and the Iran-Iraq War was restructuring military, political, and economic relationships throughout the Middle East and beyond, the situation was reversed. Now Iraq, Iran, and Libya—finding it difficult to finance essential Turkish exports—were offering as payment greater amounts of oil rather than cash. Turkey had never needed such quantities of fuel, and the three nations' cumulative debt to Turkey grew to nearly $5 billion by the late 1980s.

This economic flip-flop has had far-reaching political consequences. Whereas in the 1970s Muslim oil producers could use Turkey's overdrawn account as political leverage, in the 1980s they were asking for Turkish credit to reschedule their own enormous debts. The needs were desperately the same, but the roles were changed. Frequent defaults by these three nations, which caused several Turkish firms to declare bankruptcy and gave rise to hunger strikes by Turkish workers, cast a shadow on the countries' once strong political ties with Turkey.

Accumulating debts by Iraq, Iran, and Libya and a general reduction in its share of trade with the Muslim world reduced political pressures on Ankara, which once threatened to damage its relations with the West in general and with Israel in particular. Since the mid-1980s Turkey has been growing ever closer to Israel, a process that has included an upgrading of the level of diplomatic representation in both Ankara and Tel Aviv. The fact that Turkey was able to cut off the flow of water to Syria and Iraq for several weeks in early 1990 was a highly dramatic indication of its ability to shrug off its dependence on Muslim goodwill—the dominant feature of Turkey's foreign relations in the nearly forgotten 1970s. Indeed, in January 1992 Turkey and Israel announced the establishment of full diplomatic relations at the ambassadorial level for the first time in history.

Thus a survey of Turkey's relations with the oil countries is in effect the story of the bipolarity of the Muslim world, with the secular-democratic pole represented by Turkey, and the fundamentalist pole by Iran and Saudi Arabia. The rapprochement between Turkey and the Arab nations during the 1970s made few contributions to democratization and modernization in the Arab world, but it did draw Turkey closer to the conservatism that is the hallmark of Muslim politics and toward fundamentalism, from which Turkey had been withdrawing for most of the twentieth century. Only during the 1980s and 1990s was Turkey able to resume its Western orientation, endeavoring to become a full member of the European Common Market and later the European Union.

Islam in Modern Turkey

In the Ottoman Empire, Islam enjoyed a position of paramount importance. From its inception to the end, the empire dedicated itself to the preservation and promotion of Islam. As Bernard Lewis put it, "The Ottoman Sultans gave to the *Şeriat* [Sharia'a], the holy law of Islam, a greater degree of real efficacy than it had had in any Muslim state of high material civilization since early times. In a sense it may be said that the Ottomans were the first who really tried to make the *Şeriat* the effective law of the state."[1] In no other Muslim nation at the time were the Sharia'a so well established or the clergy so politically influential.

Nevertheless, Ottoman Turkey also had close ties with Europe; indeed, for much of its history it encompassed some parts of Europe. Thus the Ottoman elite was well aware of its country's backwardness and weakness compared with European progress. Closing the gap meant instituting far-reaching reforms, which Mustafa Kemal (later known as Atatürk) was determined to impose when he became the first president in 1923. In Kemal's view, Islam's social and political influences bore much of the blame for the erstwhile empire's backwardness. He therefore endeavored to turn Islam from a political force into merely a matter of personal belief. His reforms were, naturally enough, resisted by groups that felt threatened. These groups regarded Kemal's efforts as an attempt to impose European customs on a Muslim nation.

Nevertheless, Kemal persisted and prevailed. He saw the institution of caliphate as a link tying Turkey to the past from which he wanted to break away, and he abolished it in 1924. This was a decisive blow, soon followed by the abolition of religious schools (replaced by secular ones) and Sharia'a courts of law. The next steps were equally painful, aimed as they were against Islamic symbols: the fez and other religious garments were outlawed in 1925, and the National Assembly rejected Sharia'a in 1926. Under Turkey's new laws, polygamy and offhand divorce of women were abolished, Muslim women were allowed to marry non-Muslim men, and all adults were allowed to convert if they so wished. In April 1928 a crucial step was taken: the article in the constitution that declared Islam to be the state's official religion was dropped, and Turkey officially became a secular nation. Later Turkey dropped other remnants of its Islamic past: the Arabic alphabet was replaced by a Latinized one (1928), and Sunday replaced Friday as the mandatory rest day (1935). The school system was instructed to emphasize studies of Turkey's pre-Islamic past, tracing the roots of modern Turkey to the Hittite Empire (first century B.C.E.).

These reforms were aimed not against religion itself but against its role in politics. From that time on, the clergy no longer controlled politics, society, and culture but was restricted to the mosques. When completed, the Kemalist reforms made religion into another department of the government bureaucracy and the clergy into a group of minor officials. Despite much popular consternation, especially in the rural provinces, the clergy was helpless in the face of Kemal's enormous prestige, charisma, and powerful state machinery.

Yet the Kemalist reforms left a void that could not be filled by either the state or the vestiges of its pre-Islamic past, and the nation was still in search of spiritual and cultural inspiration and a sense of continuity and historical roots. Since the 1940s, Islam has been slowly reestablishing itself in education, society, and even politics. Even today, far from an accomplished fact, secularization is a matter of heated public debate.

The various religious orders have played a major role in the resurgence of Islam in Turkey. During Ottoman times, state religion drifted further and further from popular belief. Religious functionaries became rich, aloof, and hereditary; popular belief was focused instead on the various dervish orders. These orders have remained close to the people, and their influence remained considerable even in republican times. Although officially dismantled and outlawed in 1925, they have kept up their educational, cultural, and social activities. Since the 1960s they have been insinuating themselves into political activity as well and have set up or acquired newspapers and other mass communication media.

At present, about a dozen religious orders are active in Turkey, most of them Sunni and a few Shi'ite. The major ones object to the secular republic and wish to see it replaced by a theocratic state. For the time being, the orders' political impact is limited by their inability to support a single political party: they tend to split their support between two, three, or even more parties. But the potential is there for them to become a force to be reckoned with in Turkish politics.

The most active in politics is the Nakşibendi Order, which is also the largest. At least one government minister (Fahrettin Kurt, a former energy minister) was a member of this order, as was Korkut Özal, whose brother Turgut was Turkey's president. Also influential are the Nurcu, Süleymancı, and Kadiri Orders. Nurcu members were involved in a 1986 attempt to "reform" cadets at Turkey's Military Academy, an effort that resulted in the expulsion of about a hundred cadets.

Turkey's left-wing press has been noting with alarm the emergence of more and more activist religious groups, as well as other aspects of what it regards as Islamic reactionism. This alarm has been shared in certain influential political circles. Thus President Kenan Evren warned, in an address at Istanbul University in October 1985, against increased activities by radical Islamic organizations, calling on the universities to become bastions warding off religious fundamentalism. A few days later he warned the general public against "Communism, Fascism, and religious reaction," calling for a relentless struggle against all three. In fact, the coup d'état that brought Evren into power had been motivated largely by the military's concern over this very development. The politicians, however, were less worried. In a joint meeting of the government and the National Security Council in July 1986, attended by the president, it became clear that whereas Evren saw fundamentalism as an actual threat, Prime Minister Özal and other politicians saw it as only a potential threat. Thus, in January 1987 parliament turned down a motion to launch an official inquiry into the activities of fundamentalist groups.[2]

Is Islam a threat to Turkey, and if so, is it an actual or a potential one? It is difficult to tell. The Turkish political system is still overwhelmingly secular. Pro-Islamic groups within it constitute at the most one-fifth of the general public. Nonetheless, they are rather influential, and under certain circumstances, both domestic and international, their influence might further expand.

Religion and Party Politics

Until 1946 Turkey had a single-party system. The ruling Cumhuriyet Halk Partisi (CHP, Republican People's Party), under Kemal, was fiercely secular. It never hesitated to make use of the constitutional articles that allowed government to define anyone opposing its reforms, including on a religious basis, as reactionary subversives to be dealt with accordingly. Its policies broke the clergy's back in urban centers and strengthened the secular elites but have left a residue of bitterness in rural areas.

In 1946 Turkey became a multiparty democracy, and political attitudes toward religion changed almost overnight. Religious leaders were feeling freer to preach against secularity and to demand a greater role for Islam, and political leaders were quick to recognize the influence the clergy still enjoyed among the voting public. Religious circles were finally able to support a party of their choice and, at least in theory, to decide the outcome of elections. From the start the Demokratik Parti (DP, Democracy Party), which entered the scene in 1946, was sympathetic to religious demands and was well rewarded in terms of popular support. The ruling CHP, not to be outdone, decided it could no longer ignore religious voters. Thus in 1949 it led the move to include religious studies in school curricula and to establish a theology faculty at Ankara University, where clergy could be trained.

In the 1950s, when the DP was in power, Islam reestablished itself in Turkey's cultural life. Public calls for prayer by muezzins were again allowed, radio began broadcasting readings from the Quran and religious ceremonies, religious education was expanded to high schools, and pilgrimages to Mecca became subsidized. The number of new mosques surged and with it the number of religious publications, educational institutions, and the press. Religious orders were no longer persecuted and began moving closer to the political system. The debate over Islam's role in Turkish life became public and intense. Foreign observers were talking about the "Islamic revival" in Turkey.

In 1961, following the 1960 coup, Turkey adopted a new constitution. It still stressed the state's secular nature and rejected the use of religion for political purposes, but at the same time it provided for greater freedom of assembly, allowing—perhaps unintentionally—for the establishment of hundreds of religious organizations. Some of those organizations quickly found their way into the heart of the political system.

There were no religious political parties in Turkey until the 1970s. Some attempts were made to establish such parties in the 1950s and 1960s, but they failed either on their own or as a result of action taken against them by the government. At its peak in 1969 the Turkish Unity Party won less than 3 percent of the popular vote and disappeared soon afterward. But conditions were ripe for the appearance of a full-fledged religious party. The Adalet Partisi (AP, Justice Party), which replaced the DP in government after the 1960 coup, was becoming increasingly identified with big business, gradually losing its traditional appeal to the middle class outside major urban centers. These voters were looking for new leadership, which they found in the person of Necmettin Erbakan.

In January 1970 Erbakan established the National Order Party. Constitutionally, it could not declare itself a religious party, but its political philosophy was clear. Erbakan, a devout Muslim and a professor of mechanical engineering, moved into politics in 1969. The military government established in March 1971 banned his new party, but Erbakan was not daunted. In October 1972 he established a new party, the Milli Selamet Partisi (MSP, National Salvation Party). Less than a year later the MSP (nominally led by another, Süleyman Arif Emre, to avoid trouble with the law) faced its first challenge in the 1973 national elections and passed with flying colors. Winning 11.8 percent of total votes, translated into 48 seats in parliament, it became the third-largest party after the CHP (186 seats) and the AP (150 seats). In view of the bitter resentment between the CHP and AP, virtually no government was possible for either big party without forming a coalition with the MSP, which thus gained clout much beyond its actual weight. As soon as the elections were over, Emre resigned in favor of Erbakan, who became Turkey's first leader of national stature to wear Islamic colors.

In February 1974 Ecevit's CHP created a coalition government with Erbakan's MSP. Erbakan became deputy prime minister, and seven other members of his party became ministers. The coalition lasted less than a year because of incessant friction between its components. Among other things, Erbakan attempted to conduct his own foreign policy vis-à-vis the Muslim nations in general and Saudi Arabia in particular, as will be discussed later. It took the AP's Süleyman Demirel six months to create an alternative coalition, which again included the MSP alongside the extreme right-wing National Movement Party. Again Erbakan was deputy prime minister, and again his party received more than its equitable share of ministerial portfolios.

In the next general elections (June 1977) the MSP won only 24 seats in parliament, in part because the electoral system had been changed. But it was included in the next coalition government as well, which was headed by Ecevit. That government, however, lasted only eighteen months. In the next coalition, led by Demirel, which was removed from office by the September 1980 coup d'état, the MSP was not a member.

Nevertheless, in its six years of participation in government, the MSP achieved the most important credential it needed for survival: legitimacy as a political party of a distinctly religious hue. Because of constitutional constraints, it was unable to declare itself as such in so many words. Thus its platform could not directly call for the establishment of a religious state or for a struggle against Kemalist secularism. Instead, it emphasized Islamic virtues, the need for religious education, stronger ties with other Muslim nations, and similar issues. The party's ongoing political activity, however, was much more explicit. It fought for the institution of Friday as the weekly rest day (instead of the Kemalist Sunday), for the criminalization of offenses against Allah and Islam, for larger budgets for religious affairs, and so on. The party also fought against alcohol consumption, obscene magazines, and women in official positions wearing what it regarded to be indecent clothes.

The MSP was pushing the limits during its years in opposition, finally going beyond the pale in September 1980. Following its successful campaign to remove Foreign Minister Hayrettin Erkman from office for his pro-Western and pro-Israeli attitudes, the MSP organized a massive demonstration against Israel for its adoption of the Jerusalem Law (discussed later). The demonstrators wore traditional Islamic garments, waving green flags and chanting slogans for the reinstatement of an Islamic state. The military, Turkey's traditional bastion of secularity, was alarmed. When a generals' junta staged a coup a few days later, it cited the demonstration as one of its main reasons for doing so. General Evren, the junta-appointed president, said a few days after the coup that the event "demonstrated the scope of [religious] reaction. The nation woke up and realized the danger in all its aspects." Referring to Erbakan's address at the demonstration, Evren said: "It was said there that if our children had been kneeling down before their teachers, as is the custom in Islamic schools, they would not have become anarchists. We had not had such a school system even one hundred years ago."[3]

After the coup the MSP was outlawed, along with all other political parties, and its leaders (like theirs) were banned. In 1982 the junta drafted a new constitution, intended to amend what the generals and the public had seen as weaknesses in the 1961 one. The former constitution's too liberal nature, it was felt—specifically its broad interpretation of the tenet of freedom of association—was responsible for the anarchy of the 1970s. Under the new constitution, political parties were no longer allowed to establish youth or women's organizations or to cooperate with voluntary groups. The electoral system was changed, establishing a "blocking percentage" of 10 percent of the total popular vote; parties not attaining that percentage could not be represented in parliament. The intention was clear—to avoid splitting parliament into a multitude of small factions. At the same time, however, the 1982 constitution reintroduced religious education into all schools, presumably on the premise that since religion was going to be taught anyway, it was better that it take place under government auspices and control rather than underground. But it is doubtful whether the actual results were the intended ones.

When Turkey reverted to civilian rule in 1983, a new religious party was established, the Refah Partisi (RP, Welfare Party)—a reincarnation of the still banned MSP. It could not take part in the November 1983 election, however, since only three parties were allowed to do so. And Erbakan was still prohibited from any involvement in politics. Only in September 1987, following a plebiscite, were formerly suspended politicians allowed back, among them Erbakan. But time was too short for him to get organized before the general election held in November that year, and his new party failed to win the necessary 10 percent of the total vote.

Erbakan had changed little. He continued to attack secularity, charged Turkey's leadership with kowtowing to the Jewish lobby in the United States, led anti-Israel demonstrations, and warned against Turkey joining the European Common Market, which would make it "a province of Israel." (His subsequent career is discussed in Part 3 of this book.)

During Erbakan's forced absence from politics, many in the leaderless middle echelons of the former MSP joined the newly formed Anavatan Partisi (Motherland Party), which won a plurality in the 1983 election. The party's leader, Turgut Özal, had once been a member of the MSP and was even its nominee to parliament. He allowed the former MSP members to organize as a faction within the party, well aware that they wished to do so to promote their religious agenda.

The re-Islamization of Turkey thus gained yet another dimension in party politics. Once the MSP resumed operations under its new guise as the RP, there were two important elements in the Turkish political system pushing for a greater role for religion in public life—the RP and the well-organized and active religious faction in Anavatan. This faction derived its main support from Anavatan provincial branches, because religion has always been stronger in Turkey's rural areas. It was headed during the early years by Mehmet Keçeciler, who became Anavatan's vice chairman, and Vehbi Dincerler, who became minister of education.

Both these politicians found themselves in many political storms. In particular, almost all of Dincerler's policies produced heated public debates—he advocated prohibiting the teaching of Darwin's theory of evolution in schools, imposing "decent" sportswear on female students, prohibiting beer commercials on radio and television, including the Arabic language in high school curricula, including Islamic texts in various textbooks, dropping a series of modern words from radio and television broadcasts, and building mosques at the National Assembly compound, several government offices, and universities (some faculty members were fired for protesting against the latter proposal).

As a result of all this, during the 1980s religious education in Turkey expanded by leaps and bounds. The number of schools training imams and preachers rose from just a few in the 1950s to 717 in 1987, with 11,500 teachers and 238,000 students. In other words, about 15 percent of Turkish high school pupils (30 percent of male pupils) went to these schools, even though jobs could be found for only 2,600 out of a total of 40,000 graduates each year.

The number of religious primary and secondary schools grew rapidly as well: by the late 1980s there were 700 primary and 400 secondary schools, again compared with just a few several decades earlier.[4] Additionally, as noted earlier, religious studies became mandatory in secular schools as well. There were two theology departments in Turkey's universities at the start of the 1980s and eight at the end of the decade. Mosques were set up on twenty university campuses where none had existed before 1980. In May 1987 a student was murdered at Van University for eating during the holy month of Ramadan.

All this took place against a background of foreign involvement in Turkey's religious life. Recall that the late 1970s and early 1980s were a troublesome time for the Turkish economy, and successive governments, in a desperate quest for cheap oil, were giving in to demands from the more fundamentalist oil producers (especially Saudi Arabia) to allow them to support Islam in Turkey (as well as among Turkish workers overseas). Oil money was pouring into Islamic activities in Turkey during the 1980s, financing many of the schools and mosques described here, as well as publications, magazines, and even clergymen's salaries. (These topics are discussed in detail in the following chapters.)

The Kurdish Problem

One of the most complicated and intriguing issues affecting Turkey's relations with its Arab neighbors is the Kurdish problem, which is inextricably involved in Turkey's foreign relations as well as with its politics and the question of Turkey's national identity. According to various estimates, there are approximately 20 to 25 million Kurds worldwide, about half (10 to 12 million) in Turkey. Since 1925 the rest of the Kurds have been divided among Iran (about 25 percent), Iraq (about 15 percent), Syria, various former Soviet republics, and smaller groups consisting mainly of political refugees in Europe and the United States.

Before World War I the Kurds, who adhere to the Muslim Sunni persuasion, lived in tribal groupings of various sizes under the Ottoman and Persian Empires. After the war Britain, France, the United States, and Ottoman Turkey signed the Sèvres Agreement, which called for the establishment of a Kurdish state; but that was not to be. In 1923 Mustafa Kemal entered the Lausanne Agreement with the Western powers, which gave Turkey most of the territory originally designated for the Kurdish state. The oil-rich Mosul area was incorporated into Iraq (then under a British mandate), and the Jazirah and Kurd-Dakh Provinces were given to Syria (then under a French mandate).

In countries where the Kurds form a significant national minority, since the 1920s they have been struggling for independence or at least for the preservation of their traditions and heritage. Kurdish history during the twentieth century consisted of a series of failed revolts against the British, Turks, Iranians, and Iraqis. Some of those rebellions were brutally crushed. At times, apprehensions over Kurdish national aspirations were the only thing these countries had in common.

The main burden of the Kurdish struggle against Turkey's rule during past decades was borne by the Parti-ye Kerkaran-î Kürdistan (PKK, Kurdistan Workers' Party). The party's founder and leader, Abdullah Öcalan, lived in Syria during the 1980s and 1990s, where his party's headquarters were located.[5] The PKK started in 1978 as a Marxist-Leninist party. Its activities were financed in part by Kurdish exiles, and its guerrillas trained mainly in Lebanon's Beka'a Valley, which is effectively under Syrian control. Thus Syria has harbored the party's leader, allowed its fighting forces to train in areas it controls, and perhaps even allowed their infiltration into Turkey—a source of constant friction in Ankara's relations with Damascus.

In 1984 the PKK launched an armed struggle for Kurdistan independence. In the violence that ensued, around 30,000 lives were lost on both sides. Turkey's southeastern provinces, where most Kurds live, have been under martial law since the September 1980 military coup.

Kurdish national aspirations have been a cause of grave concern for Turkey. Any Kurdish success, within or outside Turkey, could stir unrest and undermine Ankara's authority in the southeastern part of the country. Thus the national aspirations of Turkey's Kurd minority have found almost no expression in the country's political system or cultural life. This is the result of a long-standing policy that sought to assimilate the Kurdish people: Ankara has never recognized the legitimacy of Kurdish identity. For many years the Kurds were officially called "mountain Turks." Their language was banned, as was giving Kurdish names to their children. Even the use of the words *Kurd* and *Kurdistan* was outlawed. During the 1980s the world media reported cases of forceful removal of Kurdish citizens from their homes and closures of Kurdish-operated printing presses. Punishments meted out to those convicted of Kurdish separatism were severe, occasionally giving rise to protests by Amnesty International and the European Parliament, which in May 1990 called on Turkey to recognize the political, cultural, and social rights of the Kurds.[6]

During the 1950s Turkey's Kurdish citizens had enjoyed a certain measure of liberalization, including partial right to use their language. In the 1990s they saw a guarded willingness to meet some of their demands. In January 1991 Turkey's government announced that a partial use of the Kurdish language, in both writing and public speaking, would be allowed, as would the use of the words *Kurd* and *Kurdistan* and the giving of Kurdish names. This decision was taken during the Gulf War and may have had something to do with the prevailing uncertainties regarding the fate of Iraq and its Kurdish population. Turkish newspapers reminded readers that the Turkish provinces of Mosul and Kirkuk had been taken away by the British and printed maps that depicted these areas as the Turkmen Province of Iraq.

In any event, Turkey's treatment of the Kurds has been more liberal than that of its neighbors, Iraq and Iran. Kurdish refugees escaping persecution—mainly by the Iraqi army—have often found refuge in Turkey, which gave asylum to 100,000 Kurds during the Iran-Iraq War, half of them in 1988 alone. (In one Iraqi operation against the Kurds in the town of Halabjah, which involved massive use of chemical weapons, about 15,000 lives were lost.)

In the spring of 1991, as soon as the Gulf War was over, Saddam Hussein turned to settle his accounts with Iraq's Kurdish and Shi'ite citizens. Within a few weeks more than a million Iraqi Kurds were on the run. Around 750,000 fled to Iran, and 280,000 made their way to Turkey. Another 300,000 Kurds found refuge on the Iraqi side of Turkey's border, where the United States, other Western countries, and international organizations provided them with emergency aid—including an airlift of foodstuffs and equipment. Turkey joined this effort, building access roads and supplying water. Shortly after hostilities had ended, most Kurdish refugees chose to return to their villages. Sadly, Turkey's provision of aid and shelter to hundreds of thousands of Iraqi Kurds did not end the armed clashes between PKK guerrillas and Turkish security forces.

Turkey's reexamination of its policies toward its Kurdish minority and the Kurdish issue as a whole led to several meetings during 1991 between Turkish president Turgut Özal and Iraqi Kurd leaders. In the wake of these discussions it seemed that Turkey no longer objected to the establishment of a Kurdish autonomous area in Iraq—at least not as vehemently as before. Özal must have realized that without some concessions to the Kurds, Turkey would have a difficult time fending off European criticism of its human rights record, a perennial obstacle to its European Union (EU) membership.

Nonetheless, Turkey's basic policies have remained unchanged: the country's territorial integrity rules out Kurdish independence. Turkey's leaders and diplomats have ever maintained that the Kurds in Turkey are not a minority but rather are part of the majority. To the present day, Turkey has avoided recognizing the legitimacy of Kurdish culture, heritage, and language. Clashes with Kurdish guerrillas in the southeast, almost a daily occurrence, resumed as soon as the Gulf War ended and talks with Kurdish leaders were terminated. Kurdish civilians in several southeastern towns have increasingly confronted Turkish security forces.

On 10 July 1991, 8 Kurds were fatally shot in the town of Diarbakır during the funeral of a Kurdish opposition leader. On 4 August 1991, 9 Turkish soldiers were killed in a Kurdish attack near the Turkey-Iraq border. The next day Turkey's army moved into Iraqi territory, occupying in a combined air and land effort a strip 3 miles wide and 200 miles long along the border. Ankara declared this strip a security zone, established to protect its territory from Kurdish separatists. In this action, according to Turkish officials, 700 PKK guerrillas were killed. Although Turkey and Iraq had an agreement allowing pursuit of Kurdish rebels into Iraqi territory, Ankara had failed to inform Baghdad of its intentions, and Iraq protested vehemently to Turkey and the UN. During fall 1991 and spring 1992, Turkey's army again went into action against Kurdish separatists, including launching air strikes. Kurdish guerrilla activities during that period were more intense than at any time since 1984, forcing Turkey to keep a force of 65,000 soldiers and 30,000 police along its eastern border. In the fall of 1992 Prime Minister Demirel issued a series of warnings to Syria, Iran, and Lebanon, demanding that they cease their support of Kurdish separatists.

Kurdish political rallies were often dispersed by the security forces, sometimes brutally. In March 1992 about fifty people were killed in clashes between Kurdish separatists and security forces in the southeast. The confrontation involved the Kurdish New Year celebrations (Nevruz), which take place in the early spring. The PKK called on its supporters to hold huge public celebrations, which Turkish authorities declared unlawful and took steps to prevent.

In view of Turkey's efforts to move closer to the United States, U.S. views on the Kurdish problem have become particularly significant. On 4 March 1991 the U.S. State Department issued a guarded statement that seemed to lean toward the Turkish position: the United States did not support boundary changes in the region, and it was willing to discuss with the Kurds human rights issues but not political rights. There is no doubt that this statement showed deference to Turkey's position on the Kurdish issue. The territories where most Kurds live belong to three UN members (Turkey, Iran, and Iraq), so the UN can do virtually nothing to help the Kurds realize their aspirations. And since Kurdish areas in all three countries are rich in natural resources—oil, water, chromium, iron, and uranium ores—boundary changes there could only bring about severe political and military turmoil. Thus there has been much international reluctance to make any move in that direction.

During the early 1990s, while still fighting in eastern Turkey, PKK demands became somewhat more moderate. The movement's leader, Abdullah Öcalan, tried to give the struggle a more political nature with a series of (unilateral) cease-fire declarations and a gradual reduction of the level of demands to Turkey (he was now willing to accept autonomy rather than demand an independent state). Ankara persistently ignored all such efforts by Öcalan. The Turkish government continued to regard the PKK as a terrorist organization, depicting the entire Kurdish issue as a "terror problem." There could be "no negotiations with terrorists," and Öcalan's statements and appeals remained unanswered.

This Turkish stonewalling made the PKK leadership decide to continue its military activities in eastern Turkey. Violence persisted, and the large number of casualties among Turks, civilians, and the military alike made Ankara take stronger measures against the PKK and expand its political-diplomatic efforts against countries perceived as aiding and abetting the organization. Turkish military moves into Iraqi territory in pursuit of PKK activists became routine. The struggle against the PKK came to a head in October 1998, when Turkey backed its demands that Syria deport Öcalan and cease all assistance to his forces with military buildup along its border with Syria, accompanied by explicit warlike threats. This move proved surprisingly effective. President Hafiz al-Assad—fearing escalation on the Turkish front at a time when tensions were mounting on the Israeli front and well aware of Turkey's tightening military ties with Israel, which he deemed threatening to Syria—chose to reduce tensions with Turkey.

On 20 October 1998 Turkey and Syria signed an agreement whereby the Syrians would cease all assistance to the PKK. Indirectly, they thus admitted for the first time that such assistance had been provided in the past.

From that point on, things moved quickly. Significantly, Öcalan was soon expelled from Syria, found his way to Russia, and after a few weeks was deported to Italy. The Italian government refused to extradite him to Turkey, and a wave of anti-Italian protest swept through Turkey. The protest included demonstrations all over the country, as well as boycotts—both organized and spontaneous—of Italian imports. Eventually, the Italians found a way to remove Öcalan from their territory without turning him over to the Turks. After two weeks in Italy, Öcalan began a short odyssey from one international airport to the next—always pursued by Turkish security, always asking for asylum, always being denied. On 18 February 1999 he was caught in Kenya, hiding at the residence of the Greek ambassador. He was then flown to Turkey accompanied by Turkish security. Some of Turkey's tarnished self-respect was thus restored.

Öcalan was immediately brought to trial and within a few weeks was sentenced to death for "crimes against the Turkish people." The verdict produced a massive wave of protests all over Europe. Eventually, it was decided at the highest political level to suspend Öcalan's execution pending a decision on his appeal by the European Supreme Court of Human Rights. Undoubtedly, that move had everything to do with the European Union's foreign ministers' decision in December 1999 to include Turkey among the candidates for full union membership. As of January 2001, Öcalan's fate was still hanging in the balance.

Water and Politics

Water is a decisive factor in Turkey's relations with its neighbors, particularly Syria and Iraq. Turkey is the only Middle Eastern nation with enough water resources not only to provide for its own needs but also to solve the water problems that plague nations south of its borders, such as Syria, Lebanon, Jordan, Israel, the Palestinian Authority, and even Saudi Arabia and other Gulf nations. In the various multilateral forums of the Middle East peace talks, Turkey has shown its willingness to be involved in planning and implementing projects designed to alleviate the overall water shortage in the region. For years, Turkey has believed it holds the key to solving its southern neighbors' water problems. A Turkish proposal to build a "Peace Pipeline" from southern Turkey through Syria, Lebanon, Israel, and Jordan to Saudi Arabia is still a concrete plan on the Water Committee agenda. Estimated to cost $21 billion, this project could supply these nations with up to 6 million cubic meters of water per day.

Joining in these multilateral talks on water was so important for Turkey that it agreed—after forty-four years of waiting by Israel—to upgrade the level of diplomatic representation in both countries to full ambassadorial level, Israel's condition for Turkey's participation in the talks.

As a nation of plenty amid water-starved neighbors, Turkey has frequently experienced tensions with some of the more southerly nations. The headwaters of both the Euphrates and Tigris Rivers, on which both Syria and Iraq so heavily depend, are located in Turkish territory, which makes Turkey's relations with those nations all the more sensitive.

Tensions mounted in early 1990 when Turkey stopped the flow of the Euphrates River for an entire month during the construction of the Atatürk Dam.

Turkey regards the development of its water resources in the southeast as a highly important national priority. The creation of reservoirs and the proper use of water, it believes, could solve the nation's economic problems. The main thrust of this development effort is the South-East Anatolia Project, of which the Atatürk Dam—one of the largest in the world—is the centerpiece. Work on this project began in 1983. Eventually, Turkey will have built twenty-one dams and seventeen hydroelectric power stations along the Euphrates and Tigris Rivers. When completed, the project will provide irrigation to a 30,000-square-mile area, as well as 24 billion kilowatt-hours (kWh) of energy—increasing the country's overall production of electricity by 70 percent. Through this project Ankara hopes to make Turkey's southeastern region the grain basket of the entire Middle East.

Understandably, the same project is a cause of great concern in Syria and Iraq, because it would give Turkey full control over a flow of water amounting to 32 billion cubic meters. Having already shown it can completely cut off this flow, Turkey has strengthened its bargaining position in its complex relationships with its southern neighbors. Although Ankara has repeatedly reassured them that it would never use water as a weapon, Damascus and Baghdad remain suspicious. Meanwhile, the Syrians and Iraqis are well aware that filling up the Atatürk Dam reservoir will perforce involve a reduction of the flow downstream.

A lingering problem in this tripartite water conflict is the absence of any explicit agreement among these nations on water allocation.[7] This is a difficult problem, because no internationally agreed principles exist for water sharing among riparian states. At present, the international norm emphasizes the right of each riparian state to "a fair share" of the river's water. Moreover, the headwater state is expected to inform downstream nations in advance of any changes it wants to make in flow control, and it cannot implement such changes without their consent.

Turkey's position is fairly simple: the Euphrates is a Turkish river. Understandably, this is not the view taken by Syria and Iraq, which regard it as an international watercourse requiring an international agreement for the allocation of its water in accordance with the provisions of international law. Pending a formal allocation agreement between Turkey and its neighbors, the World Bank and other international agencies refuse to assist Turkey in financing the South-East Anatolia Project.

Unlike Iraq, which also enjoys the water of the Tigris, Syria is almost completely dependent on the Euphrates River. The river is Syria's main source of water (80 percent of its total potential) for both agricultural development and power. Of the 32 billion cubic meters flowing down the Euphrates each year, Syria uses 15 billion. Neither agricultural nor industrial development could take place without that amount of water. Hence any threat to Syria's share of the Euphrates water is literally vital to its survival.

Iraq, too, is heavily invested in development plans that could come to naught if its share of Euphrates water were significantly reduced, because the Tigris does not provide an adequate alternative. Tigris water is used for irrigation elsewhere, and the quality of that water is continually deteriorating. Iraq has demanded from Turkey and Syria an annual amount of 18 billion cubic meters per year. That is more than the total amount flowing out of Turkey to Syria, to say nothing of the amount actually reaching Iraq. Obviously, then, there is no way to provide all three nations with the amount of water they require. The Euphrates therefore promises to remain a serious bone of contention for years to come.

Occasionally, this conflict over water has become entangled with the Kurdish issue and other regional problems. In January 1990, following a series of news reports about missile development by Turkey's Arab neighbors, several Turkish leaders departed from their usually restrained treatment of such sensitive issues. Minister of State Kamran İnan, for instance, told the newspaper *Cumhuriyet* (12 January 1990), "They have missiles; we have water." President Turgut Özal said his country would not agree to sharing Euphrates water, only to arrangements regulating its flow, because this was a technical rather than a political or legal problem.[8] In early 1992 Turkey began to insist on receiving an overall plan for Euphrates water use from Syria and Iraq as a precondition for any binding decisions on its part regarding flow control. Turkey's U.S. ambassador, Nüzhet Kandemir, pointed out that Turkey had no intention of giving away its water, just as the Arabs were not giving away their oil.[9]

At the start of the twenty-first century, Turkey's surplus water resources still provide it with significant economic and political advantages, particularly in the wake of the drought in areas to its south during the late 1990s. Most countries there (especially Jordan) experienced severe water shortages and are looking for both immediate and longer-range solutions. The complex political problems dividing Turkey and Syria have put Turgut Özal's Peace Pipeline plan on the back burner, but several alternatives have surfaced in recent years, such as the use of huge floating plastic containers called jellyfish, an underwater pipeline (which has proven impractical so far), or tankers previously used to carry oil that would be refitted for water haulage.

This last possibility came to the fore in early 2000, when Turkey completed the construction in Manavgat, 50 kilometers east of Antalya, of offshore water terminals for tankers. Israel, Jordan, and the Palestinian Authority are all exploring with Turkey the possibility of importing water in this way.

Notes

1. Bernard Lewis, *The Emergence of Modern Turkey* (Oxford: Oxford University Press, 1968), pp. 13–14.

2. *Briefing* magazine, 19 January 1987.

3. *Pulse,* 15–17 September 1980.

4. Schools whose curricula and rules conformed to the teachings of Islam, as distinct from the schools training clergy discussed earlier.

5. The Turkish government claimed for many years, and Syria persistently denied, that Öcalan was holed up there. This was officially proven to be the case in October 1998. Shortly afterward, he was expelled to Russia.

6. Bruce R. Kuniholm, "Turkey and the West," *Foreign Affairs* 70, no. 2 (spring 1991).

7. There is an arrangement (as distinct from a formal agreement) between Turkey and Syria whereby Turkey agreed to allow an average flow of 500 cubic meters per second into Syria.

8. *Ha'aretz* (Hebrew), 24 June 1990.

9. Public lecture at Emory University, Atlanta, Georgia, 19 March 1992.

PART ONE

TURKEY'S DEPENDENCE ON OIL

2

THE ECONOMICS
OF ENERGY IN TURKEY

Turkey became highly dependent on fuel imports just before the energy crisis broke out in December 1973, because it reached the takeoff stage in its economic development in the early 1970s.[1] Like some nations in southern Europe and Latin America, Turkey found it difficult to curb fuel imports in the first few years after the outbreak of the crisis. Any slowdown would have run against the grain of its development plans, and heavy financial investments already made would have gone down the drain.

Turkey's Energy Sources

Compared with other nations in a similar stage of accelerated industrialization, however, the energy crisis impact on Turkey's economy and standard of living was exceedingly severe. Adjusting the country's energy policies to the new circumstances took unusually long, and during the lengthy period of fuel shortages (1977–1980) repercussions were felt in every part of Turkey's economy, bringing far-reaching realignments in both the social fabric and the political system.

At the time, Turkey's per capita energy consumption was low by Western standards but rather high when compared with the underdeveloped Third World. Oil accounted for half of Turkey's energy sources in the period 1973–1982. The rest included noncommercial energy (wood and dung)—about a quarter of the total—as well as lignite, coal, and waterpower. Thus local resources—noncommercial fuel, lignite, most of the coal, and part of the oil—provided a little more than half of total energy consumption. Turkey had to import the rest from abroad— mostly oil, but also small quantities of coal, liquefied gas, and electricity.[2]

Between 1969 and 1977 the share of oil (particularly imported oil) as part of Turkey's energy resources increased, and coal and noncommercial sources decreased in importance. During that same period oil imports increased from 2.9 to 11.7 million tons per year.

In 1977 oil consumption peaked at 53 percent of Turkey's energy resources. In contrast with the Western industrial nations, the steep increase in imported oil's share among energy resources continued during the period 1974–1977. In part, this was a result of the high growth rate of Turkey's economy, but in part it was caused by a decline in domestic oil production and by policies that kept energy products (especially fuel) at considerably lower prices than those in other European nations. Domestic political instability made it difficult for Turkey's various governments to bring fuel prices up to real levels; public outcry would have been overwhelming. At the same time, Turkey did little to develop any substitutes for oil or even to launch a serious energy-saving campaign.

As a result of these factors, between 1974 and 1977 the demand for energy increased at a rate of 9 percent per year, and the financial expenditure on fuel imports rose sharply (see Figure 2.1). Despite this rapid increase in its oil bill and serious financing difficulties notwithstanding, Turkey was unable to reduce imported oil's share of its energy sources during the late 1970s and early 1980s. Domestic oil production continued to decline during the period 1977–1981, and energy consumption trends continued to increase in 1979–1981.

As we shall see, Turkey's main problem during the period 1977–1980 was how to pay for oil rather than where to get it. Thus we ought to look at Turkey's expenditure on oil imports as part of its overall import costs and the share of the oil bill relative to overall export revenues. This examination should highlight the burden oil imports imposed on Turkey's economy. If we look at the official statistics published by both the Turkish government and the International Monetary Fund, the following trends appear:

Figure 2.1   Turkey's Oil Import Expenditures, 1971–1984

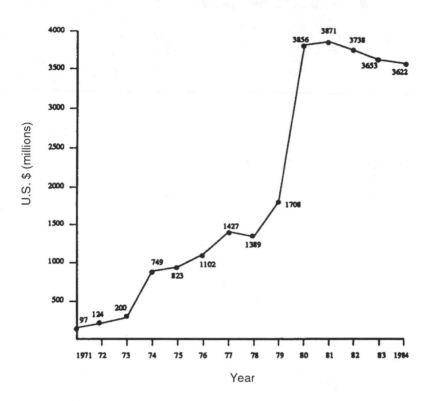

Sources: Turkish State Institute of Statistics, *Statistical Yearbook of Turkey* (Ankara: 1972–1985); Energy Ministry of the Republic of Turkey, *The Energy Resources of Turkey* (Ankara: 1984).

1. Between 1973 and 1980, the oil bill's share of overall import expenditure (cost, insurance, and freight [CIF]) increased from 9.3 to 49.5 percent (see Figure 2.2).

2. The expenditure on oil imports as a percentage of Turkey's total export revenues increased from 17 percent in 1973 to 132 percent in 1980; that is, in 1980 the oil import expenditure was one-third greater than the total of Turkish export earnings (see Figure 2.3).

Figures 2.2 and 2.3 illustrate these trends. Thus it can be seen that 1980 was the most difficult year for Turkey to cope with its oil bill.

**Figure 2.2   Share of Oil Imports in Turkey's Total Imports, 1970–1983
(as a percentage of CIF imports)**

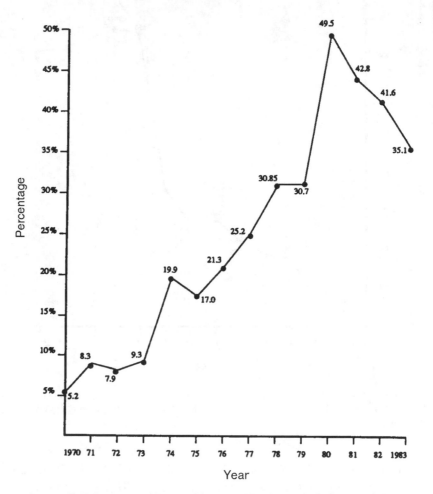

Sources: IMF, *International Financial Statistics* (Washington, D.C.: January 1982), p. 400,
(1984), p. 446; Turkish State Institute of Statistics, *Statistical Yearbook of Turkey* (Ankara:
1975–1984).

Since then, the situation has eased significantly. In 1984, for instance,
oil import expenditure amounted to only 50 percent of total export revenues—almost the same as 1974, immediately after the outbreak of the
oil crisis. The year 1980 also stands out as a black year in the comparison between oil import expenditures and Turkey's foreign currency
reserves (see Figure 2.4).

Figure 2.3   Oil Imports as a Share of Total Exports, 1973–1984

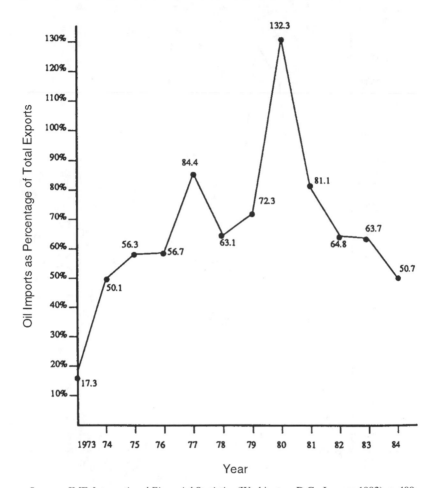

Sources: IMF, *International Financial Statistics* (Washington, D.C.: January 1982), p. 400, (1984), p. 446. IMF, *Turkey—Recent Economic Developments* (Washington, D.C.: August 1981).

Although Turkey's foreign currency reserves showed nominal recovery during the period 1977–1981, having reached an average level in excess of $1 billion in 1980,[3] we must take into account the huge increase in Turkey's oil bill during that period. The net result is that in 1980 Turkey's foreign currency reserves were less able to cope with oil import expenditures than they had been in 1973: whereas in 1973 Turkey's reserves could cover its oil bill for that year nine times over, in

32

Figure 2.4 Turkey's Foreign Currency Reserves, 1970–1983

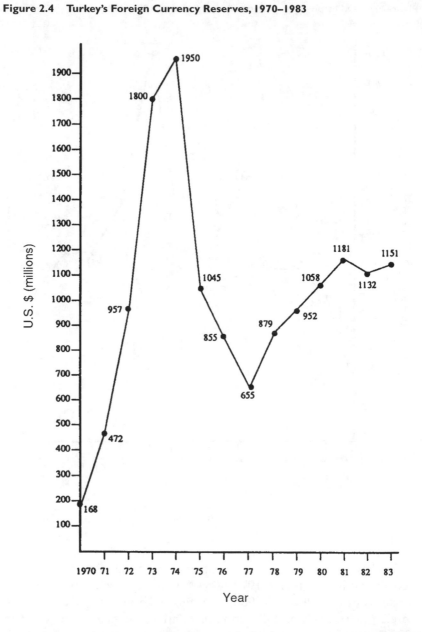

Source: IMF, *International Financial Statistics* (Washington, D.C.: January 1982, March 1984).

Note: Foreign exchange annual averages, excluding gold.

1977 they could cover oil imports for only 169 days and in 1980 for just 100 days.

This increased dependence on oil imports during the 1970s grievously affected Turkey's balance of payments. In 1973 that balance (including unilateral transfers) was positive $515 million. Increased oil import expenditures, coupled with decreased unilateral transfer payments because of an economic recession in the West as a whole, created a $3.88 billion deficit in Turkey's trade balance within four years, resulting in an overall deficit of $3 billion in balance of payments.

During this time Turkey was almost completely dependent on Muslim countries for its oil (see Figure 2.5).[4] This fact had several political repercussions for Turkey's foreign relations (see Chapters 4, 5, and 8). The three most important oil suppliers during the 1975–1984 period were Iraq, Libya, and Iran; Iraq was Turkey's chief oil supplier during most of the 1970s. Supply of Saudi oil to Turkey ceased almost completely during the years 1975–1980.

Comparative Dependence on Oil:
Turkey and Other Oil Importers

It is interesting to compare Turkey's oil dependence with that of other importers. For comparison, I chose three nations that imported vast quantities of oil from the Middle East during the period 1971–1981—Brazil, India, and Japan—to see how they financed their oil bills.

Clearly, Turkey's dependence on Muslim producers was the highest among these four nations (see Figure 2.6). With the exception of 1979 (when financing problems forced Turkey to turn to the free market for distillate imports), during the second half of the 1970s Brazil, India, and Japan reduced their dependence on Muslim oil, whereas Turkey proved unable to do so. Indeed, Turkey remained highly, if not completely, dependent on oil from Muslim sources even in the early 1980s. In fact, Turkey barely looked for oil sources outside the Middle East in the first place.

To appreciate these nations' ability to finance oil imports, we must look first at the burden their respective oil bills imposed on their national economies. For all these nations, the oil bill's share of total import expenditure increased in the years 1974–1979. In financial terms it rose from 22 to 25 percent in 1974 to 35–38 percent of total import costs in 1979, adversely affecting the ability to import other goods.

Figure 2.5   Turkey's Oil Imports by Source, 1972–1984

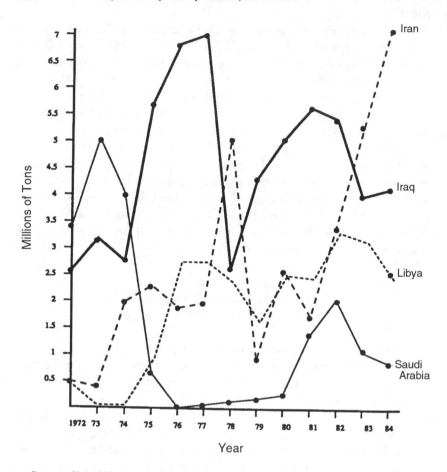

Sources: United Nations, *World Energy Supplies* (New York: 1977, 1979, 1980, 1982); Turkey's Ministry of Energy publications (1982–1984); Sevinc Carlson, *Oil Haves and Have-Nots—Turkey as a Case Study* (September 1972).

Comparing the size of the oil bill with total export revenues for these four nations (Figure 2.7), we can readily see that Turkey's financing ability was by far the weakest. In terms of export revenues, Turkey's oil bill was unusually high. Again, the seriousness of Turkey's position was most obvious in 1980, although the country had more difficulty financing its oil imports than Brazil, India, or Japan during the whole period between 1974 and 1986.

A similar picture emerges from a comparison of Turkey's financing

Figure 2.6   India, Brazil, Japan, and Turkey: Oil Imports from Muslim Countries as a Percentage of Total Oil Imports, 1971–1981

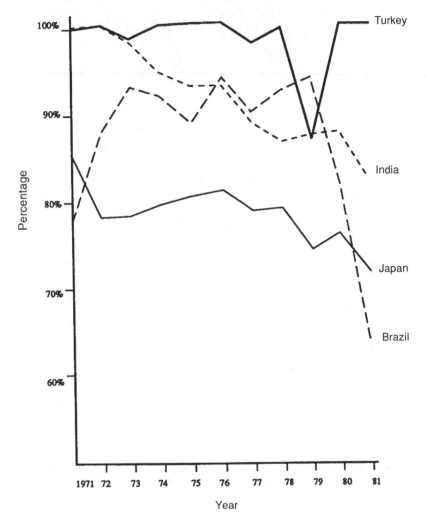

Source: United Nations, *Energy Statistics Yearbook* (New York: United Nations, 1970–1984).

problems with those of the same group of nations. Furthermore, an examination of Turkey's ability to finance its total imports (not only oil) through export revenues (Figure 2.8) underlines Turkey's economic weakness during the 1970s.

In conclusion, a comparison of oil import financing ability among

Figure 2.7   India, Brazil, Japan, and Turkey: Oil Imports as a Percentage of Total Exports, 1973–1986

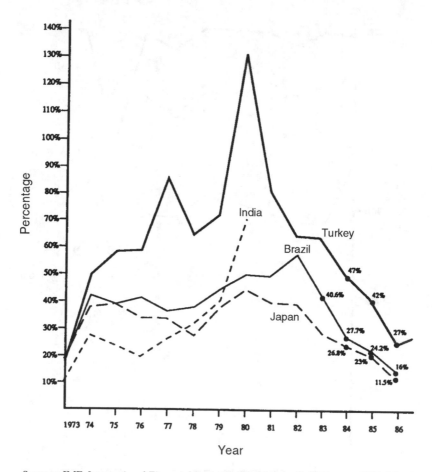

Sources: IMF, *International Financial Statistics* (Washington, D.C.: January 1982, March 1984); Economist Intelligence Unit, *Turkey,* Country Report No. 3 (London: 1989).

Turkey, India, Brazil, and Japan shows conclusively that Turkey was in a very weak position during the second half of the 1970s. Japan's position was fairly sound, despite the huge nominal size of its oil bill, whereas Brazil and India also suffered from financing difficulties as a result of a massive debt burden (Brazil) or increasing difficulties with the trade balance (India). Both countries, however, were in a better position than Turkey.

Figure 2.8   India, Brazil, Japan, and Turkey: Ability to Finance Total Imports Through Export Revenues, 1972–1984

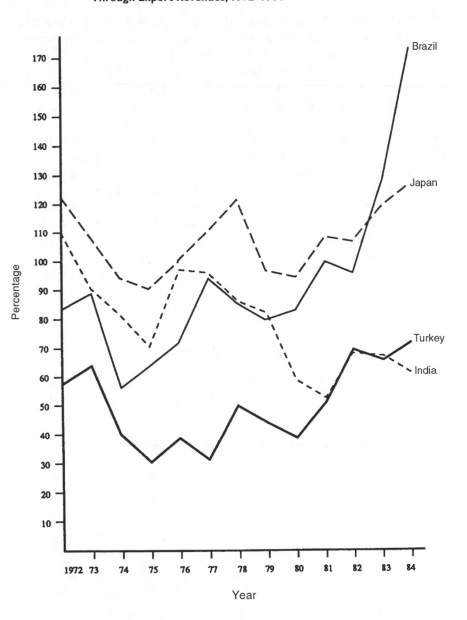

Source: IMF, Direction of Trade (Washington, D.C.: 1979, 1985).

Stockpile Levels

The level of oil stockpiles maintained by an importer reflects its determination to invest in preparedness for possible shortages in the future,
as well as its estimate of the reliability of its oil suppliers. In the
absence of full and reliable data on the oil stockpiles maintained by
Brazil and India in the period 1974–1982, I compared Turkey's stockpiles with those kept by Japan, Spain, and Portugal during the same
period, because these nations were also highly dependent on Muslim oil
imports (see Table 2.1). Examination shows that Turkey's stockpile levels in the early 1980s remained at about forty days' consumption, much
the same as in 1974. Despite the increasing uncertainties brought about
by the oil crises since 1973 and ignoring International Energy Agency
(IEA) guidelines, which placed stockpile recommended levels at ninety
days' consumption, no significant change occurred in Turkey's stockpile levels during the period. Japan, Spain, and Portugal, on the other
hand, increased their stockpiles by 70 to 100 percent between 1974 and
1982. Thus in the early 1980s Turkey's stockpile levels were significantly lower than those of the other nations. This fact, as we shall see,
increased Turkey's vulnerability to pressure from the oil producers.

Turkey as an Oil Carrier

Alongside Syria, Egypt, and Jordan, Turkey was one of the nations
referred to as "the oil carriers group" in the 1970s, positioned as they

Table 2.1   Oil Stockpile Levels, 1974–1982 (days of consumption, annual average)

	Turkey	Japan	Spain	Portugal
1974	41.0	67.0	78.0	61.0
1975	39.0	74.0	90.5	59.0
1976	38.0	74.0	87.0	70.0
1977	36.0	77.0	113.5	83.0
1978	n.a.	77.0	103.0	70.0
1979	29.0	81.0	102.0	73.0
1980	34.0	102.0	113.5	123.0
1981	43.0	107.5	128.0	132.0
1982	37.0	111.0	131.0	123.0

Source: Calculated from oil stockpile quantities published by Central Intelligence Agency
(CIA), *International Energy Statistical Review* (Washington, D.C.: CIA, 1977, 1979, 1980,
1983).
Note: n.a. = not applicable.

were between oil production regions in the East and consumption centers in the West.[5] In view of Turkey's geographic position, intense discussions had begun in the late 1960s and early 1970s about the possibility of building gas and oil pipelines from the Gulf to the Mediterranean Sea through Turkish territory. After years of futile negotiations with Iran (discussed later), in 1973 Turkey decided to enter a pipeline agreement with Iraq. The agreement, signed on 27 August 1973 following brief negotiations, called for a 981-kilometer (km) pipeline to be built between the Kirkuk area in northern Iraq and Dörtyol in the Gulf of İskenderun. The Kirkuk-Dörtyol Pipeline Agreement, accompanied by an oil supply contract, had far-reaching consequences for Turkey's energy position in the late 1970s and throughout the 1980s.

According to this twenty-year agreement, the Kirkuk-Dörtyol pipeline was to have a carrying capacity of 35 million tons per year, which could be increased in the future. Iraq undertook to run through the pipeline at least 15 million tons per year, paying Turkey thirty-five cents for each barrel of oil in passage. Each party was financially and technically responsible for building the pipeline in its own territory (341 km in Iraq, 640 km in Turkey).

Although the agreement was reached fairly quickly, Turkey took its time ratifying it. The outbreak of the energy crisis in October 1973 made the Turks think twice about some of the agreement's provisions, which no longer fit the new realities. The main source of frustration was the fact that passage fees were fixed rather than linked to crude oil prices, originally expected to fluctuate with international market trends. The sharp increase in oil prices in late 1973 and early 1974 applied to Iraqi oil as well, and now Turkey had to pay that price while passage fees remained at their original level.[6] Ankara was also concerned that it may not be able to bear the heavy financial burden involved in building its part of the pipeline. Decisionmakers in Ankara feared Turkey would pay for most of the pipeline while Iraq would reap most of the project's benefits. In addition, the pipeline was to run through Kurdish territory in both Iraq and Turkey, and there was the additional concern about political instabilities and possible disruptions by Kurdish rebels. Therefore Ankara made new demands to Baghdad to increase passage fees for oil flowing through the pipeline, as well as for a credit line for 10 million tons of oil to be purchased during the years 1974–1976 to enhance Turkey's ability to pay for the pipeline's construction. Both demands were rejected, and Turkey's opposition to ratification became stronger.

Supporters of the agreement argued that it was precisely because of the new circumstances created by the energy crisis that Turkey needed a reliable source of oil just across its borders. They expected the Kirkuk-Dörtyol pipeline to provide Turkey, as a member of the oil carriers group, with a measure of influence over oil politics in the Middle East and claimed investments would be repaid within eight years at the most. Withdrawing from the project would mean Turkey would give up a source of income amounting to $100 million per year.

On 28 January 1975, about eighteen months after the pipeline agreement was signed, Turkey's senate ratified it. From that time on, construction proceeded swiftly, and on 3 and 4 January 1977 the pipeline was inaugurated in two separate ceremonies, one in Iraq and one in Turkey. Turkish prime minister Süleyman Demirel and Iraqi vice president Taha Ma'aruf were in attendance. This was a boon to Iraq, which was now able to resume the flow of oil to the Mediterranean, previously interrupted because of a conflict with Syria in April 1976. Iraq's dependence on Syria was dramatically reduced, whereas Turkey's importance as a convenient route for Iraqi oil exports increased. Damascus was furious with Iraq because of this decision to give Turkey preference as its main outlet for oil, which robbed Syria of $165 million a year in passage fees. The Syrians construed Iraq's decision as a conspiracy to weaken Syria economically, politically, and militarily.

During 1977 the Kirkuk-Dörtyol pipeline carried around 8 million tons of oil, but in December of that year Iraq decided to stop using it because of an unpaid Turkish oil bill amounting to $230 million. The pipeline's operations were suspended until August 1978, when Turkey and Iraq reached a new oil agreement.

Besides Turkey's financing problems, additional factors adversely affected the pipeline's operations in 1977 and 1978: (1) the costs involved in maritime oil transportation were greatly reduced, (2) technical problems involving power supplied in eastern Turkey disrupted the pipeline's automatic control system, (3) terrorist activities in the area included sabotage of the pipeline itself,[7] and (4) the Iraqi marketing system allowed Turkey to load and ship oil using only tankers approved in advance by Iraq. Only in 1980 did the Kirkuk-Dörtyol pipeline begin to carry amounts of oil approaching its full capacity, 35 million tons per year. Barring the first few weeks of the Iran-Iraq War, those years were the first golden age for the pipeline. Prime Minister Demirel, at a press conference on 10 March 1980, called the Iraqi-Turkish pipeline "Turkey's black gold belt."

Beginning in 1979, the pipeline proved economically beneficial to both countries: delivery time for Gulf oil to Turkey was cut from forty-five days (by sea) to two days, for a long period (when Gulf ports were blockaded because of the Iran-Iraq War) the pipeline was the only outlet to the west for Iraqi oil, and Turkey has enjoyed greater revenues, having insisted—from a position of strength now—that passage fees be significantly increased. Ankara has used the pipeline as an important bargaining chip in its relations with Baghdad, offsetting to an extent Iraq's great advantage as Turkey's chief oil supplier.

The pipeline's economic and political advantages to both countries led Turkey and Iraq to decide to increase its carrying capacity from 35 to 50 million tons as of 1984 and to sign (in April 1985) another agreement for a parallel pipeline between the same terminals. This additional pipeline became operational by mid-1987. Between them, both pipelines can carry 1.6 million barrels per day (80 million tons per year). Operating at full capacity, they could secure annual revenues of $350 million per year in passage fees for Turkey.

The economic boycott imposed on Iraq following its invasion of Kuwait in 1990 was therefore a serious economic blow to Turkey. The pipeline was closed down following firm U.S. demands, and Turkey thus lost huge potential revenues during most of the 1990s (the pipeline resumed partial operations only in mid-1999).

Turkey's Energy Policies

The problems Turkey encountered following the outbreak of the energy crisis in late 1973 were similar to those affecting many other nations at a stage of accelerated economic growth in the early 1970s. This category included most southern European nations (e.g., Greece, Portugal, Spain, and Yugoslavia) and a number of nations in Latin America and Asia. The common denominator for them all was their inability to curb domestic demand for energy or to increase domestic production.

The main difference between these nations and Turkey lay in the severity of the impact of the crisis. Turkey's energy sector, and subsequently its economy as a whole, was grievously hit. The prolonged fuel shortage Turkey experienced between December 1977 and March 1980 was unusual in comparison with other nations at a similar stage of development. Turkey's balance-of-payments problems were unusually severe, as we have seen, and consequently industrial production declined while unemployment and inflation were on the increase. By the late 1970s Turkey's economic crisis was far more serious than that in most other nations at a similar position in the early 1970s in terms of their ability to cope with the energy crisis.

A well-thought-out, firm energy policy during the years 1974–1977, emphasizing energy saving and resource development, would have significantly reduced the deterioration of Turkey's economic situation. But Turkey barely tried to manage the crisis during those years. The prices of energy products on the domestic market were not increased frequently enough, and no funds were allocated for the development of domestic and alternative energy resources. Political instability made politicians ever fearful of voters' reaction to unpopular but necessary policy measures. Also, because of good neighborly relations and Muslim affinities, Turkish leaders had hoped for much more generosity than was shown by Muslim oil producers when Turkey found it difficult to pay its oil bills.

Because of its accelerated economic growth rate—about 8 percent per year—Turkey had to keep increasing its oil imports despite the steep rises in costs. The extra money needed to pay for its mounting oil bill came from "burning" foreign currency reserves and increasing borrowing from abroad, mostly in the form of short-term loans. Turkey's short-term bank loans amounted to $145 million by the end of 1974, rising to $1 billion by the end of 1975 and reaching $1.8 billion by the end of 1976. By 1979 Turkey ranked with Zaire and Peru as one of the world's countries least likely to repay its debts.

A series of newspaper articles by Professor Mükerrem Hiç strongly criticized the Turkish government for serious failures in developing fuel, hydroelectric, and thermal energy resources.[8] Among other things, Hiç alleged that during the premiership of Bülent Ecevit, Turkey's reliance on Soviet and eastern European technology and liquidation of private investments in the energy sector, both ideologically motivated, had greatly slowed the development of energy resources in Turkey. This view largely reflected the position taken by Adalet Partisi (the Justice Party), which called for reliance on private entrepreneurship and Western technology for the development of Turkey's energy sector—policies that were out of favor during most of the 1970s.

Even Energy Minister Kamran İnan admitted at the third Energy Congress in Ankara on 16 November 1977, "While all nations of the world took steps to save fuel, we took the opposite direction, encouraging oil consumption."[9] Gradual remedial steps taken in 1979 and 1980 (especially increases in fuel prices, adoption of energy-saving measures, and increased investments in the energy sector) came too slowly and too late and thus were unable to prevent serious disruptions of the economy and hence of Turkey's life in general. Policies desperately needed in the years 1974–1979 in such areas as resource allocation and domestic fuel prices were not adopted until 1980. By then, the global energy crisis was resolving itself, so these new policies were largely irrelevant.

Meanwhile, the deterioration of the balance-of-payments position triggered a chain reaction that in final account proved disastrous for the Turkish economy. During the late 1970s Turkey's inability to pay for oil imports sufficient to meet demand caused a partial shutdown of productive sectors and shortages of many raw materials and consumer goods. In vain, Ankara hoped for special consideration by Organization of Petroleum Exporting Countries (OPEC) members, let alone for petrodollar investments. By the end of the decade Turkey realized that the solution to its problems must be more radical than an overall reform of the energy sector. Turkey's export destinations had to be realigned, and perforce its foreign policy toward the Middle East had to be adjusted to meet the new conditions created by the energy crisis.

The 1979–1980 Fuel Shortages

Until the fall of 1977, Turkey had barely experienced any real fuel shortages. From that time until March 1980, however, Turkey went through a period of intermittent fuel shortages that reached its peak in late 1979 and early 1980. During this period fuel stockpiles deteriorated to the level of a few days' consumption (even zero levels for several distillates), and the government was struggling to obtain the missing fuels.

Even though fuel shortages had become endemic by the end of 1977, Turkey's government took no significant measures to rectify the situation until early 1979. At that time Energy Minister Deniz Baykal proposed a series of steps intended to cope with the problem, such as restricting the use of private vehicles, increasing use of railways in preference to motor cars, and using coal instead of oil products for heating. In March 1979 Turkey's National Fuel Company and the governors of the various districts issued regulations whereby preference in fuel distribution was given to commercial and public vehicles over private ones. Some gas stations were required to sell only to public transport vehicles. On 1 April 1979 a government regulation banned the import of cars with more than four cylinders, and in early July that same year fuel rationing began in the big cities: private vehicles were allowed to use up to 80 liters of fuel a month, compared with 500 to 900 liters for public transportation vehicles.

Despite these rationing measures Turkey's oil situation continued to deteriorate, and by October 1979 shortages were widespread all over the country. Eventually, the government was forced to announce an emergency plan for fuel allocation, prioritized by province and sector. Agriculture and several manufacturing industries headed the priority list.

There were several reasons for the severe oil shortages that hit Turkey in late 1979: (1) severe financing problems, which became even more acute following the sharp increase in oil prices during 1979; (2) an inability to meet obligations to export goods (particularly cereals) under a series of barter agreements signed with Iran, Libya, and Iraq in 1978; (3) refusal by Saudi Arabia, Kuwait, and the United Arab Republic to provide Turkey with credit lines for oil imports, despite repeated pleas; and (4) fuel hoarding all over Turkey, triggered by public concern over future fuel supplies.

Between November 1979 and February 1980, these fuel shortages hit all sectors of the economy hard, affecting most of the population as well. Blackouts were numerous and ever lengthening, heating oil became scarce even on the black market, and food and coal supplies to the cities, desperately needed in the harsh winter, were disrupted by fuel shortages. Gas supplies suffered when it was decided to lower pressure in gas lines, and cooking became impossible. The only rays of hope were banner newspaper headlines announcing that huge oil supplies were on their way to Turkey. These were mostly false; they were planted, or so it seems, to raise morale and silence public outcries.

The damage caused by the fuel shortages was widespread. The education system, unable to provide heating for classrooms, was completely shut down; the health system was nearly paralyzed by a lack of pharmaceuticals and basic medical equipment; manufacturing industries operated at about 20 percent capacity for lack of raw materials and spare parts; unemployment was increasing; and inflation reached a rate of 80 to 100 percent per year. The economic slump was inevitably accompanied by social unrest, including rampant violence and political terror.[10]

On 17 January 1980 a Western diplomat in Istanbul sent his home office a letter describing in detail living conditions in Turkey. Concluding a thorough discussion of shortages of fuel and other essential commodities, the diplomat wrote: "As a result . . . there is an increase in the use of electricity for heating, which in turn overburdens the city's power mains, although these were inadequate to begin with. Blackouts in various parts of the city are now longer and more frequent, sometimes lasting for 36 consecutive hours; this in the absence of any other source of heating, since trucks are not running for lack of fuel." Inevitably, this unhealthy situation led to many diseases, particularly among children, yet pharmacies were short of even the most basic drugs.

Demirel, who was prime minister at the time, described this period several years later:

> In November 1979, when government had no money to buy electric bulbs, we became a *petroleum yok* [no fuel] nation. As soon as we came into power, I made a great effort to overcome those shortages. Every month, I brought in twice the amount of fuel needed in ordinary times. There was a psychology of shortage, and the public were hoarding. We flooded the market with huge quantities of fuel, and everything disappeared immediately. It was only by March 1980 that we were able to overcome fuel shortages and excessive hoarding.[11]

And indeed, on 20 March 1980, Energy Minister Esat Giritlioğlu officially announced that the fuel shortage was over. The government revoked all rationing regulations imposed during 1979 and allowed the National Petroleum Company and foreign fuel companies to sell fuel products to all comers without limitation. During the spring the situation improved greatly, and on 1 May 1980 Demirel was able to announce proudly on Radio Ankara that Turkey was starting off that month with a thirty-day fuel stockpile—"for the first time in fifteen or eighteen months." Later that month Turkey was reported to have resumed the export of petroleum distillates to the tune of $30–$35 million.

Turkey and the International Energy Agency

Turkey's effort to cope with the energy crisis by aligning itself with Western consumer nations was accompanied by an intense public debate, focusing from 1974 onward on Turkey's membership in the International Energy Agency. This debate clearly brought out Turkey's uncomfortable and ambiguous position between the Muslim oil producers' group on the one hand and Western consumers on the other.

The idea to create a cartel of oil consumers to stand up to OPEC and prepare for a possible future embargo was conceived by the U.S. secretary of state, Henry Kissinger, in early 1974. Representatives from twelve Western nations spent the summer of that year hammering out a draft proposal for the creation of the IEA.[12] Subsequently, it was decided that membership would be open to all members of the Organization for Economic Cooperation and Development (OECD), which would implement IEA requirements on issues such as oil stockpiles, emergency oil distribution, joint energy research, and so on. The draft proposal defined the IEA as an "autonomous agency within OECD."

On 16 November 1974 the international press published for the first time the list of sixteen IEA members, which included Turkey.[13] This was a surprise, because the question of Turkey's membership had never been discussed by either the domestic or the international press. Furthermore, Prime Minister Ecevit and his second in command, Necmettin Erbakan, were rather sympathetic to the oil-producing nations (their attitudes are further discussed later). When OECD finance ministers convened prior to IEA's inauguration, the Turkish representative, Deniz Baykal, rejected the idea that Turkey should become a founding member, claiming such a posture would run against the grain of his government's basic policy guidelines.

Yet Turkey later announced that it had become an IEA member. The Arab countries expressed displeasure, and a widespread public debate ensued. How should Turkey relate to IEA and Western alignment vis-à-vis the energy crisis? Arab diplomats in Ankara wasted no time in reminding Turkey that the agency was established to combat the oil suppliers, and they exerted a great deal of pressure against Turkey joining. Baghdad reacted furiously, making fresh and compelling demands on Turkey to pay its accumulating debts.

The debate was grounded in economics as much as politics. Many supporters of Ecevit's Cumhuriyet Halk Partisi (Republican People's Party), as well as of Erbakan's religious Milli Selamet Partisi (National Salvation Party), saw Turkey's move as an affront to the oil producers, with which Turkey was endeavoring to develop "special relationships." The conservative Adalet Partisi, on the other hand, saw the move as a "natural" one for a Western consumer nation hard hit by the sharp increase in oil prices. Remarkably, both Prime Minister Ecevit and his foreign minister, Turan Güneş, had supported the decision to join, at least at the time the proposal was made; they realized only later how strong the opposition was to this move within their own Republican Party. In fact, the decisive force in leading Turkey into the IEA was the Foreign Office: the decision was made during the transition period between two cabinets, Ecevit's and Sadi Irmak's, at a twilight time in which the Foreign Office was able to exert strong influence.

In the energy sector, opposition to Turkey's IEA membership was based primarily on economic considerations. It was led by Irmak's energy minister, Erhan Isıl, and the professionals in charge of fuel procurement in both his office and the National Petroleum Company. They were chagrined by the Foreign Office's failure to consult with them before making its decision to join the IEA. Apart from immediate difficulties with oil acquisition, these professionals were also concerned about conflicts of interest between Turkey and most other IEA members, which were highly industrialized consumer nations.

The Turkish press largely confined itself to reporting the IEA debate without taking sides. An exception was the left-wing *Cumhuriyet*, which saw the decision to join as fundamentally wrong and criticized the Foreign Office for "butting into" the IEA without invitation. *Cumhuriyet* took up the public struggle against Turkey's participation in the IEA as part of the more general campaign launched by most U.S. papers against a background of strains in U.S.-Turkey relations.

The Arguments in Favor of IEA Membership

The position taken by those who supported Turkey's membership in the IEA was best laid out in a report prepared by Turkey's representatives at OECD under instructions from the Foreign Office in Ankara.[14] The report highlighted the IEA's raison d'être: to conduct a peaceful defensive struggle against the potentially serious consequences of the oil price hikes. The agency was described as "an economic North Atlantic Treaty Organization," and the report stated that one of its roles should be to strengthen the position of the international oil companies, which was greatly diminished following OPEC's unilateral moves. The report concluded that the IEA was essential for safeguarding the interests of both OECD members and developing nations faced with the energy crisis.

"We wanted to use IEA know-how and financing in the development of oil substitutes and in the emergency oil distribution plan," said Energy Minister Selahattin Kılıç. As for concern over the oil producers' reaction to Turkey's decision and the prospect of losing its right to "cheap oil," Kılıç explained: "We haven't thought about the producers in this respect. By the time we joined IEA, we had already lost the 'special prices' for Iraqi oil we had had in 1974 [discussed later]. And as far as supplies were concerned, we knew Iraq and Libya will keep on selling to us in the future, too."[15]

On 28 May 1978, Kılıç told a press conference in Paris, while attending an IEA conference there, that "oil imports are draining Turkey's lifeblood."[16] He did not think Turkey's membership was prejudicial to its relations with the Arab nations. On the contrary, by joining this and other international forums, Turkey could provide a bridge between producers and consumers, thanks to its geographical position and historical traditions.

Yet another aspect of the pro-IEA faction's position was highlighted by Serbülent Bingöl, energy minister in 1980–1981: "IEA was not intended to fight oil producers; rather, it was created to protect the interests of the consumers. Why should the producers have a right to unite and the consumers not have it?"[17]

The Arguments Against IEA Membership

Opposition to IEA membership among Turkish politicians was even more vocal, and their arguments were more diverse. The leaders of the opposition, who effectively prevented ratification of Turkey's participation in the IEA, were Erhan Isıl, who became energy minister immediately after the decision to join had been made, and Deputy Prime Minister Necmettin Erbakan, whose Salvation Party was a member of all of Turkey's various coalition governments during the period 1974–1978. The reasons for their opposition were completely different. Isıl based his objections on substantial economic arguments, whereas Erbakan's motives were political in nature.

Isıl was not impressed with the "Western solidarity" argument for Turkey's participation because of the economics of the situation: "Western industrialized nations could protect themselves by raising the prices of their industries' sophisticated products. Turkey's level of development was a far cry from Germany's or Japan's, and thus our interests were different too. We had an energy deficit, and we were not in the same position to close it as other European nations [were]." And indeed, some IEA requirements—such as a ninety-day stockpile, approval of energy policies of member nations, provision of information in advance of oil transactions, and similar issues—were inconsistent with Turkey's economic needs.[18]

The oil producers' displeasure with Turkey's participation in the IEA, said Isıl, figured little in his decision to oppose the move: "I did not object because we were about to ask for or receive cheap oil. From the start it was my view that we would be unable to acquire cheap oil from the producers, IEA notwithstanding."

For Turkey's third Oil Congress, which convened in Ankara in 1976, Isıl wrote an article elaborating his position on Turkey and the IEA.[19] According to his view, the decisionmaking mechanisms designed for the IEA made it impossible for minor consumers such as Turkey to achieve the majority required (60 percent) for the agency's emergency plan to come into operation. The financial burden involved in meeting the agency's huge stockpile requirement would damage Turkey's development plans. Besides, IEA members with limited financial resources and low industrial levels would be unable to work out their energy programs on their own. Further, Isıl regarded the IEA as a "countercartel" put up by the industrialized bloc against OPEC, something he thought would only exacerbate the conflict between producers and consumers.

Erbakan based his opposition to Turkey's IEA membership on his Salvation Party's political and economic ideology. In fact, he objected to the very idea of the IEA, not only to Turkey's participation in the agency. He regarded it as a Western anti-Muslim body created specifically to combat and thwart OPEC. He believed Turkey should have nothing to do with any agency created to combat OPEC. This view was shared by others outside the Salvation Party, particularly after U.S. secretary of state Kissinger mentioned in January 1975 a possible use of force by the Western bloc against the oil producers.

Turkey's IEA Seesaw

The disagreement between IEA supporters and detractors in Turkey produced an ambiguous attitude that persisted until Turkey finally ratified its IEA participation in 1981. Turkey's decision to join the IEA had been made by the Ecevit government, which lost power soon afterward. The strong position against joining the IEA, taken by both the energy minister in the next cabinet and the Salvation Party, made Turkey's position in the IEA rather uncertain. Prime Minister Demirel, who wished to have Turkey's IEA membership ratified as soon as he assumed power in March 1975, nevertheless preferred to keep his coalition government intact. Faced with Erbakan's opposition, he was left with no other choice than to comment that the IEA was not a good enough reason to bring down the government. His energy minister, Kamran İnan, chose not to participate in the IEA energy ministers' summit in Paris in 1977, also because of Erbakan's position: "I felt that I could not go to Paris without a mandate to ratify the agreement."[20]

Yet despite strong objections by the Cumhuriyet Halk Partisi—a member of the opposition in 1975–1977—and the public campaign led by *Cumhuriyet* against the IEA, the Justice Party—the mainstay of Demirel's coalition government—as well as senior officials in the Foreign and Energy Offices made continual efforts to bring Turkey closer to the IEA. There was a legal problem, because Turkey had not yet ratified its decision to join the agency, but Demirel's aides were able to create good working relationships with the IEA. In June 1975 Finance Minister Yılmaz Ergenekon even decided that Turkey would join the IEA Financial Assistance Fund without consulting with Deputy Prime Minister Erbakan, who as chairman of the Ministerial Committee on Economic Affairs had the final say in the matter.

A change of government in January 1978 made Bülent Ecevit prime minister once again. His new coalition, with Deniz Baykal as energy minister, was even more hostile to the IEA than its predecessor and was more fearful of the oil producers' attitude. Turkey's ties with the IEA weakened, and rumors were spread about possible withdrawal. Only in late 1979, when Ecevit's cabinet was replaced by a minority government led by Demirel, was Turkey finally able to ratify its IEA membership. Early in 1980 the government decided to do so, and a year later, in April 1981, Turkey became a full IEA member, participating in the agency's emergency programs.

The complexity of attitudes taken by various political circles in Turkey toward the International Energy Agency, as well as Turkey's relationship with the agency, was characteristic of many other issues in Turkey's foreign relations at the time, some of which are discussed later. Frequently during the 1970s Turkey found itself caught between energy producers and consumers or, in a broader sense, between the industrialized north and the developing south.

The IEA debate reflected deep divisions inside Turkey regarding its relationship with the Muslim world. IEA supporters have maintained that complete reliance on Muslim oil producers was dangerous for Turkey. They saw a policy based on such reliance as potentially harmful should those relationships ever reach a point of crisis because of some bilateral problem or in view of the endemic instability of Middle Eastern politics in general. The other camp felt Turkey had no real choice in the matter other than to trust the Muslim oil producers. Some regarded the Muslim world as Turkey's "natural" sphere; others were anti-Western as a matter of principle. The end result was a policy see-saw, as Turkey was trying to have the best of both worlds.

In retrospect, Turkey's decision to join the IEA did not harm its relations with the oil producers. At the same time, it afforded no real protection in terms of oil supplies, even at a time when Turkey desperately needed such protection.

Notes

1. *Fuel* includes both crude petroleum and its distillates.

2. Turkish State Institute of Statistics, *Annual Foreign Trade Statistics* (Ankara: 1975–1984); Energy Ministry of the Republic of Turkey, *The Energy Resources of Turkey* (Ankara: 1984).

3. Compared with an average of $659 million in 1977.

4. UN data on Turkey's oil imports from Iran, particularly for 1974 and 1975, were much higher than the figures published by Turkish sources. Apparently they included oil purchased from Iran through third-party brokers.

5. See Benjamin Shwadran, *Middle East Issues and Problems* (Cambridge, Mass.: Harvard University Press, 1977), pp. 35–77.

6. The August 1973 agreement stipulated that Iraq would sell Turkey oil at $2.59 per barrel. No sooner was the agreement signed than the Iraqis raised the barrel price to $5.11. Following the oil embargo, in line with OPEC policy, they jacked the price up again to $17–$18 per barrel.

7. In July 1977 the pipeline was hit by guerrilla activities for the first time near Mardin (400 km east of Dörtyol). In the same area the pipeline was hit again in October 1978. Such acts of sabotage continued into the 1980s. According to Turkish security sources, the guerrillas hitting the pipeline were aided and abetted by Syria in planning and materiel.

8. Mükerrem Hiç, "The Question of Foreign and Private Capital in

Turkey" and "Economic Policies Pursued by Turkey," *Orient* 21 (1980) and 23 (1982), respectively.

9. In *Pulse,* 17 November 1977.

10. Turkey's hardships at this time were widely reported in the international press. See, for instance, *New York Times,* 13 January 1980; *Briefing,* 19 December 1979; and all Turkish dailies during this period.

11. Interview with the author, Ankara, 20 May 1985.

12. The United States, Canada, Japan, Norway, and the members of the European Common Market, barring France.

13. Eleven of the twelve nations listed in note 12 (Norway opted out), plus Austria, Spain, Sweden, Switzerland, and Turkey.

14. See *Cumhuriyet,* 19 December 1974.

15. Interview with the author, Ankara, 17 May 1985.

16. *Pulse,* 29 May 1978.

17. Interview with the author, Istanbul, 7 May 1985.

18. Interview with the author, 22 May 1985.

19. Erhan Isıl, "International Energy Programme and International Energy Agency," Third Turkish Petroleum Congress (Ankara: 1977), pp. 199–200.

20. Interview with the author, Istanbul, 9 May 1985.

3

THE DIPLOMACY OF CHEAP OIL

Turkey's rapprochement with the Arab world had begun a few years before the energy crisis. Therefore Turkish leaders expected, once the crisis loomed large, to receive preferential terms in their deals with Arab oil producers. Between March 1974 and November 1977, however, at a time when Turkey had yet to feel the full impact of the oil crisis, its efforts to obtain cheap oil from Iraq and Libya were only partially successful. Similar efforts directed at Saudi Arabia and Iran came to nothing.

Reliance on Iraq and Libya

Oil Agreements with Iraq

Even before the outbreak of the energy crisis in October 1973, Turkey decided to make Iraq its chief oil supplier. In August 1973 the government, then headed by Naim Talu, signed a long-term oil agreement whereby Iraq would be Turkey's main supplier until the mid-1980s. The agreement (which has never been fully realized) called for the export of 6 million tons of Iraqi crude to Turkey in 1977–1979, to be increased to 12 million tons in 1980 and to 14 million tons in 1983. As soon as the oil crisis started, however, a dispute broke out between the two countries about the price of the oil. The agreement stated clearly that the price paid by Turkey for Iraqi oil would be the Organization of Petroleum Exporting Countries' (OPEC) posted price. In January 1974, however, Turkey launched an intensive effort to have its imported oil price reduced. During the early stages of renegotiation Turkey stressed such emotional issues as good neighborly relations and common heritage. More realistically, other issues have become involved in the negotiations, such as the Kurdish question, Iraqi water supplies, and the joint pipeline. In those early times, Turkish representatives explicitly demanded special treatment for Turkey.

The Turkish delegation that went to Baghdad in January 1974 was headed by Ambassador Oğuz Gökmen, at the time in charge of economic affairs at the Foreign Office. In this initial effort to cope with the new realities following OPEC's decision to quadruple the price of oil at a single stroke, Turkey was content to put an official rather than a politician in charge of negotiations, reflecting its naive belief that its special treatment plea would be readily accepted by Iraq.

Ambassador Gökmen kept negotiating with Iraqi representatives through February 1974. Press reports spoke of hard bargaining and even confrontation between the two delegations, including a Turkish threat to look elsewhere for oil supplies and a reminder of the geographical location of the sources of the Tigris and Euphrates Rivers.

When a new government came into power in Turkey under Bülent Ecevit, the pace of negotiations was accelerated. The Turkish press spoke of a forthcoming agreement whereby Iraq would sell oil to Turkey during the next three months for the posted price, $12.50 per barrel, but would actually get only $11.50 per barrel, keeping the discount a secret because of Iraq's commitment to OPEC. The fact that this "secret" agreement reached the press, however, made the position of Turkey's negotiators rather awkward.

Once the matter of price was settled, Turkey's newspapers raised two essential questions: Was the price Turkey had obtained really the best one possible? And what did Turkey give in exchange? The first question was discussed extensively in the sensationalist *Günaydin* and *Hürriyet* (both running an anti-Iraqi campaign), and it seems possible that government circles were feeding them information. On 19 March 1974 *Günaydin* wrote that while Iraq was selling oil to Turkey for $11.00 per barrel, France was paying $8.00 per barrel for its Iraqi oil, and other European countries were paying $8.00 to $8.50. *Hürriyet* went further, claiming on 1 and 2 April that Iraq was reneging on its August 1973 agreement with Turkey, which specified a price of $2.80

per barrel, and had also refused to charge current oil imports against Turkey's future revenues from the Kirkuk-Dörtyol pipeline.

Turkey's Foreign Office reacted immediately with an unequivocal denial. Iraq was not reneging on its agreement with Turkey, Iraqi oil was coming in on schedule, the agreement remained in effect, and there was no Iraqi attempt to change its terms. Although the August 1973 agreement had indeed specified a price of $2.80 per barrel, it also included periodic price revisions, so *Hürriyet*'s accusations were not grounded in real facts.

These charges by the press did reflect, however, a strongly felt sense of frustration in Turkey as its unfounded hopes for special oil prices were crushed, and they created unease in relations between the two countries. Still, everything seemed to have been straightened out when an Iraqi delegation came to Ankara to discuss oil matters. Its chief, Ibrahim al-Wakil, made a statement on 9 April 1974 to the effect that "Turkey's oil price is the lowest one agreed upon by Iraq with any country, including France. Iraq does not sell oil for $8.50 per barrel to anyone."[1] He went on to say that recently Iraq had sold Turkey oil at a preferential price compared with other oil prices in the Mediterranean area. Strangely, this amounted to an admission that Iraq *was* undercutting OPEC prices in its deals with Turkey.

The media also speculated on Turkey's quid pro quo for the Iraqi goodwill. There was general agreement that Turkey had agreed to release more water for both Iraq and Syria. There were claims that allowing more water to flow down the Euphrates would slow the rate of filling the reservoir behind the Keban Hydroelectric Dam, intended to go online by June 1974. Such a delay would wreak havoc on Turkey's power supply, with grievous ramifications for the entire economy. The Foreign Office denied that the price of Iraqi oil was connected in any way with Euphrates water release: "Filling up the Keban reservoir, Syrian and Iraqi water needs, and the oil Turkey imports from Iraq are distinct issues that should be dealt with separately."[2]

Public attitudes toward Iraq became friendlier in the wake of these disarming statements. Iraq's consent to sell Turkey oil below OPEC prices was favorably compared with Saudi Arabia's rude refusal of a request made in late April 1974 for preferential treatment in oil deals (discussed later). Iraq was becoming even more popular because of the assistance it had provided to Turkey, as had Libya, in the Cyprus war.

On 2 July 1974 Iraqi vice president Saddam Hussein went to Ankara to meet with Prime Minister Ecevit. It was agreed that Iraq would provide Turkey with 4–5 million tons of oil a year on credit and would make a loan to cover the costs of the Turkish part of the Kirkuk-Dörtyol pipeline. Turkey promised to cooperate with Iraq on Euphrates water issues and in guarding the pipeline on both sides of the border. Addressing parliament the next day, Ecevit noted that the agreement resolved both oil and water problems with Iraq and created "a new atmosphere" in relations between the two nations.

Oil Agreement with Libya

Libya's support of Turkey in the 1974 Cyprus war marked a turning point in relations between the two countries. During late 1974 there were numerous exchanges of political visitors; in particular, relations were cemented between the Republican People's Party, the mainstay of the ruling coalition until September 1974, and Libya's ruling party, the Arab Socialist Union. This process culminated on 5 January 1975 in a trade agreement between Turkey and Libya.

The centerpiece of this agreement for Turkey was an oil deal for 3 million tons of crude petroleum and 200,000 tons of diesel fuel per year. Such quantities were unprecedented as far as Libyan oil supplies to Turkey were concerned. During the period 1969–1972, Turkey imported from Libya relatively small amounts of crude, between 250 thousand and 500 thousand tons per year. In 1973 this amount declined to 30,000 tons, and in 1974 no oil was imported from Libya to Turkey.[3]

Besides economics, the Turkey-Libya agreement had a military aspect as well. Economically, it dealt with permits for Turkish nationals to work in Libya, outlined joint industrial projects, called for the establishment of a joint bank, and spoke of starting a shipping line; militarily, Turkey was to train Libyan servicemen, whereas Libya would finance Turkish purchases of aircraft and spares.

Even before the agreement was signed, the Turkish press had begun speculating on the "special terms" given by Libya for Turkish oil purchases. Both Turkish and international newspapers figured that Libyan oil was sold to Turkey at 8 percent below OPEC prices.[4] Libyan oil was of high quality, which gave Turkey an invisible advantage of around U.S. 80 cents per barrel. Information now available seems to indicate that overall, the special terms given to Turkey amounted to a 20 percent discount off the current market price.

Even without accurate information it is reasonable to suppose these terms were the most favorable Turkey could have obtained at the time. This was a significant relief to Ankara, as Turkey's government was struggling to obtain oil in large quantities and on good terms. The Libyan oil agreement brought about a major change in Turkey's oil supplies. Having signed it, Turkey secured approximately 80 percent of its total oil imports: 5 million tons from Iraq plus 3.2 million tons from Libya.

Thus efforts to secure oil supplies from Saudi Arabia and Iran became less frenetic in 1975 and 1976. On the other side, however, Turkey became almost completely dependent on Iraq and Libya for its oil—a dependence that created major difficulties later when Ankara was facing the problem of how to finance its oil imports.

Efforts to Prevent Price Hikes

It seems that Turkey kept importing Iraqi oil at lower-than-OPEC prices at least until mid-1975. At that time a "mishap" occurred that made Iraq far less generous toward Turkey from 1976 on. Selahattin Kılıç, energy minister at the time, told the story:

> Our National Petroleum Company was then getting Iraqi oil at $10.50 per barrel. This was lower than the OPEC official price. In June 1975 I stood up in parliament to defend our energy policies. Because of the political pressures put up by the Republican People's Party, who were very critical of our policies, I was forced to reveal this discount in public on that occasion. This revelation lost us the opportunity to get Iraqi oil for such a price. After we'd disclosed the price I was told by 'Adnan Hussein Hamadani, the head of the Iraqi Planning Institute, that the Iraqis were unable to go on selling us oil for this price, since OPEC countries were monitoring their oil prices. Later on we made more efforts to obtain cheap oil from Iraq, but in vain.[5]

It seems the scope of Turkish-Iraqi negotiations during 1976 was wider than before. In early February 1976 Turkey's foreign minister, İhsan Sabri Çağlayangil, came to Baghdad to sign the oil agreement for the next year. Upon his return he announced that the Iraqis has agreed to proceed with the sale of 5 million tons of oil to Turkey in 1976 as well. The Iraqis did demand a higher price, but at the same time they were willing to extend their credit line for Turkish oil imports. Turkey therefore still regarded Baghdad's terms as "preferential."[6] The Turkish press believed that in view of this Iraqi credit, Iraq was actually selling oil to

Turkey at a discount of 20 percent below OPEC prices[7]—much the same preferential treatment as Turkey was receiving from Libya. Oil imports from Iraq were one of the main topics discussed during a visit to Baghdad by Turkish president Fahri S. Korutürk in late April 1976.

Despite all efforts by Foreign Minister Çağlayangil and President Korutürk, however, Iraq did raise the price of oil, even though the international oil market had reached a state of oversupply and price hikes had become rare. Still, this nominal price hike was eroded during the second half of 1976 in view of the oversupply situation.

The Deterioration of Preferential Treatment

In early March 1977 Turkish-Iraqi relations took a turn for the worse because of oil prices. The Turks had found out that even though the expected completion of the Kirkuk-Dörtyol pipeline was supposed to lower the cost of haulage by about 70 cents a barrel, Iraq was still demanding a price of $13.48 per barrel—higher than the cost of oil shipped in from the Gulf.

Disagreements over the price delayed the start of pipeline operations and brought a number of Turkish delegations to Baghdad. All efforts failed, however, despite the calm that prevailed in the international oil market. Iraq's insistence on the official price in its dealings with Turkey had to do with two extraneous problems: (1) Turkey was increasing its utilization of the Euphrates River and was planning future water projects, a source of grave concern for Baghdad; and (2) Turkey's foreign currency problems had deteriorated to the point that Turkish payments were in arrears beyond the three-month credit agreed to by Iraq.

Iraq's bargaining position grew stronger still in April 1977 when it was announced that Çağlayangil had failed in his efforts to secure oil from Saudi Arabia, Kuwait, and the United Arab Emirates (discussed later). Turkey's protestations against the method of calculating the costs of haulage now became the subject of tough and lengthy negotiations, to the degree that in early 1977 the terms of Turkish oil purchases from Libya became more generous than those provided by Iraq.

Concern over Iraqi demands gradually found its way into the Turkish press. On 5 April 1977 Energy Minister Kılıç had to admit that this prolonged dispute was preventing the flow of Iraqi oil in the now completed pipeline. As of May 1977 Turkey was forced to pay the full international price for Iraqi oil. Faced with a mounting debt to Iraq and unable to secure oil under better terms from Saudi Arabia or Kuwait, Turkey had to give up the preferential treatment it had received until then just to make oil flow through the new pipeline.

The decision for Turkey to continue to rely on Iraq as its chief oil supplier had economic and political ramifications beyond the question of price. As Kılıç explained later:

> In making certain decisions about Iraqi oil supplies, we were forcing the hands of the energy ministers who were to come after me. According to our constitution, certain government actions are subject to supervision by the Supreme Court. The government has to propose its decisions on issues of special importance to this body, which would then confirm or dismiss them. At a certain stage, we asked the court to approve our reliance on Iraq for oil. We explained to them our geographical situation and the facts created by the Kirkuk-Dörtyol pipeline and were given their approval for having Iraq as the supplier of choice for Turkish oil imports.[8]

This decision and the principles underlying it in fact made Iraq Turkey's chief oil supplier for the 1980s as well.

Failure to Diversify

Efforts in the Gulf

Turkey's first attempt to obtain oil from Saudi Arabia under preferential terms took place in January 1974 when Naim Talu's transition government was in power. It was a strange move, largely indicative of naive Turkish expectations regarding possible oil deals with Muslim producing nations immediately after the 1973 embargo.

Lütfi Doğan, the official in charge of religious affairs in the prime minister's office, was invited by the Saudis to make a pilgrimage to Mecca. The Turkish Foreign Office decided to use the opportunity of his visit to open a secret channel to explore the possibility of procuring cheap oil. Doğan was not even briefed in advance of his visit; the request was forwarded to him through Turkey's embassy in Jeddah after he had already left. Upon his return Doğan passed on to the Foreign Office both his impressions and the Saudis' demands. He then told the press that during his mission, which he himself described as "secret," he had met with King Feisal and received the impression that the king was willing to give Turkey preferential treatment—namely, a lower price for oil; at the same time, Saudi Arabia had demands of its own. Still, the entire episode was an unmitigated failure for which both Doğan and the Foreign Office bore the blame. In fact, the Saudis refused to discuss oil matters with Doğan, insisting that meaningful discussions could only take place with a high-ranking delegation headed by a cabinet minister or a presidential representative.

Still, Doğan's mention of Saudi demands launched an intensive dialogue between the press and official Turkish spokespersons about a possible Saudi demand that Turkey sever its diplomatic relations with Israel. Foreign Office officials denied such rumors: "Until now, no Arab country has made such a demand of Turkey in exchange for oil supply, nor was there the slightest hint in this matter."[9] Foreign Office sources confirmed that since the outbreak of the oil crisis in October 1973, Arab nations had indicated, unofficially and even officially, that they would be happy to see all Muslim states sever relations with Israel—a thinly veiled reference to Turkey and Iran. Yet "even though our government understands their position, the appropriate answers were given." As for rumors about Saudi pressure on Ankara, "Turkey has not even begun negotiating the purchase of oil from this country."[10] The Saudi ambassador in Ankara also denied the rumor about Saudi pressure to cut off relations with Israel as a condition for cheaper oil, stating that Saudi Arabia would never interfere in Turkey's internal affairs.[11]

Erbakan Visits Saudi Arabia

During March 1974 the Turkish press and officialdom were optimistic about Turkey's chances to reach a satisfactory agreement on Saudi oil. The Saudi ambassador in Ankara told *Yankı* magazine that his country had "2.5 million tons earmarked for Turkey."[12] Turkish optimism was based largely on the assumption that Muslim nations such as Libya and Saudi Arabia would want to strengthen the position of Necmettin Erbakan, the leader of the Muslim Salvation Party, to indicate to the Turkish public that Riyadh would deal more generously with a Turkish government that relied more heavily on Islamic elements in parliament.[13]

Thus Erbakan was sent to Saudi Arabia as head of a large delegation to discuss oil. But his mission had become shrouded in uncertainties even before he left Ankara. The day before Erbakan's departure, *Milliyet* wrote that representatives of the Turkish National Petroleum

Company (TPAO), recently back from Saudi Arabia, had been told that for the time being the Saudis had no extra oil to sell and that the possibility of granting Turkey a $1 billion credit was "nonexistent."

It was feared in government circles that Turkey would be required to pay a political price and betray Mustafa Kemal Atatürk's ideals for the sake of Saudi oil. Bearing in mind that Erbakan's Salvation Party publicly supported a break with Israel, government officials in Ankara were concerned about Erbakan's mission. A meeting of the National Security Council, convened on 23 April 1974, decided to instruct his delegation to explore in Saudi Arabia the possibilities of developing economic ties "with no strings attached" and "without prejudice to the basic principles underlying Turkish foreign policy."[14] Erbakan was also instructed to be cautious in his efforts to develop cultural ties (which he had sought to expand) and was prevented from signing a student exchange agreement. Thus his visit became limited to the commercial-economic sphere, and in the final analysis this made it impossible for Erbakan to obtain a cheap oil agreement. The entire affair reflected the prevailing attitude of Turkish decisionmakers in the spring of 1974: they were still hoping cheaper oil could be obtained from Saudi Arabia without any political or cultural concessions. They thought it was possible to convince the Saudis that Turkey was important to the oil shaikhdoms by virtue of its geopolitical position as a buffer blocking the spread of communism southward and that this was reason enough to give it preferential treatment in the supply of oil.

Although the National Security Council had defined his visit as "nonpolitical," Erbakan began to make anti-Israeli and pro-Palestinian statements as soon as he arrived at the airport. He emphasized that Israeli withdrawal was a precondition for peace in the Middle East and defined Jerusalem as a "Muslim city" whose position could not be changed by any fait accompli.

Upon his return to Ankara on 2 May, Erbakan declared that Saudi Arabia had promised to give sympathetic consideration to Turkey's oil supply requests, yet it was plain that he had come back empty-handed. Saudi Oil Minister Ahman Zaki Yamani had explained to him that Saudi oil was completely sold until 1976 and that Turkey could not be given any special price until then. Even when the delegation had persuaded Saudi officials to agree in principle to Turkish purchase deals through multinational corporations, the verbal agreement was withdrawn as soon as it was requested in writing. Time and again, Erbakan was told "we'll discuss this later."

Public criticism of the mission was increasing. The press pointed out that the Saudis were not impressed by Erbakan's personality or by his huge delegation. Some newspapers derided his Islamic gestures toward his Saudi hosts, particularly Yamani, who it seemed did not think highly of them. The deputy prime minister's extremist statements about the situation in the Middle East proved unhelpful to Turkey in its effort to obtain oil under favorable or even normal conditions.

For fear of causing an open breach with the oil producers, criticism was mainly leveled not at Saudi Arabia but rather at the leader of the Turkish delegation. While pointing a few arrows at the hosts, the Turkish press mostly dealt with their guest, pouring both derision and anger on him. *Hürriyet*'s correspondent, who had accompanied the delegation, demanded outright that Erbakan be fired, claiming he had "acted on his own, consulting no-one. . . . He made a nuisance of himself with his hosts, and because of him, the whole delegation looked ridiculous."[15] Another commentator, *Milliyet*'s Metin Toker, wrote that a coalition of which Erbakan was a member would last no more than "two *takkas*" (headgear worn by worshippers in a mosque) and demanded that the president, the prime minister, and parliament intervene and curb Erbakan's "mischief."[16]

Other commentators ridiculed the political expectations Erbakan had developed prior to his visit.[17] They claimed the deputy prime minister had been led by pro-Islamic circles to believe King Feisal felt that to bring about the unity of the Muslim world a charismatic leader was needed, one who was already the leader of a highly developed country with a strong military. Saudi Arabia did not fit that description, so the nation most suited to lead Islam was none other than Turkey, and that was a historic opportunity Erbakan could not afford to miss. He had also pinned his hopes on the fact that some members of the Saudi court were of Turkish descent and would lend him a helping hand.

Foreign Minister Turan Güneş escaped criticism by not joining the delegation to Riyadh in the first place; he had calculated that the resolution taken by the Muslim summit at Lahore in early March 1974 regarding sales of cheap oil to Islamic nations was no more than wishful thinking. In his view, Islam had nothing to do with the price of oil, as he informed members of the Senate in his report on the summit, which he had attended as Turkey's representative.[18]

The bitter failure of Erbakan's mission gradually came home to Turkish politicians, bringing about a fundamental change in expectations. Again it was Güneş who expressed these new notions in an article published in early May in the monthly *Özgür İnsan:*

> We must not be overly optimistic, letting our fancy lead us into expectations regarding our relationships with the Islamic oil nations. No one has any intention of letting anyone have cheap oil. Arab producers are careful even when they stand to gain political advantages for themselves. We must not hope that these nations will give us preference because of our historical ties with them, when our offer is less profitable to them than some other country's.

Parliament also severely criticized Erbakan. Senator Kamran İnan dubbed his trip "miserable," adding that "the spiritual principles sacrificed during this visit exceed in value all the oil wells in the world."[19] Later, in an interview in Ankara with the author on 9 May 1985, İnan said: "The whole visit was humiliating. Erbakan first asked for a billion dollar loan. When the Saudis refused, he went down to 250 million and kept going down. He invited Yamani to pray with the Turkish delegation, but Yamani went on strolling on the veranda, ridiculing him. Eventually he got a cultural agreement, which was a useless piece of paper, an insult to the Turkish nation, running against the grain of Turkey's constitution."

Interviewed by the author on 16 May 1986 in Ankara, a minister in Ecevit's cabinet who asked to remain anonymous listed several causes for the failure of Erbakan's visit to Saudi Arabia:

> 1. The Turkish Islamic extremists had a naive attitude toward Saudi Arabia. The Saudis had no time for Turkey. It was just business for them. They knew how to conduct a negotiation, and [they] stood by their own interests. Erbakan's attitude was different, completely inappropriate for dealing with the Saudis.
> 2. There was a fundamental difference between the case of Iran and Iraq and the Saudi case. With these two [former] nations, we had to maintain good relations, and the same goes for them, because we were close neighbors and had common problems in our relationships. Saudi Arabia, in contrast, wanted cash for oil and couldn't care less about [what] its relationship with Turkey looked like. With the Saudis there was no depth to our relationship, unlike the case of Iran and Iraq; there was no reciprocity.
> 3. Unlike our relationship with Libya, the Saudis didn't care

whether whoever purchased oil from them sympathized with the Arab world or the developing world. Ideology and overall view of the world, which played a major role in oil decisionmaking with the Libyans, has had no role at all to play in the marketing of Saudi oil.

4. Doing business with the Saudis is a fine and strange art. It requires flexibility and special maneuvering capabilities, which we didn't have. The Saudis preferred not to get into a situation where they had to sit [at] a table with official representatives and discuss oil deals. Many governments were forced therefore to negotiate through middlemen.

All in all, Erbakan's visit revealed to the Turks the shape of a bleak reality. In early May 1974, eight months after the embargo had been imposed, Turkey finally got the message that had been there from the start: it too, like any other consumer, would have to pay the full price for oil purchased from "sister nations." Other Turkish officials from all parts of the political spectrum would try to obtain oil from the Saudis during the later 1970s. All would receive the same answer Erbakan got: no direct deals; no special treatment.

Foreign Minister Visits the Gulf States

Following this humiliating and frustrating visit to Saudi Arabia, Turkey relaxed its efforts to obtain Saudi oil for some time. The oil deals entered into with Iraq and Libya in late 1974 and early 1975 made the need to search for oil sources in the Gulf less pressing. A few attempts were made to discuss Saudi oil after May 1974—all of them on the technical level, all of them unfruitful. Not until March 1977 did a senior Turkish leader go to the Gulf; this time it was Foreign Minister İhsan Sabri Çağlayangil.[20]

His trip took place against a background of growing foreign exchange difficulties and a shortage of oil, which was becoming increasingly felt in everyday life. On 25 March 1977 Turkey's gold and foreign currency reserves hit rock bottom: $645 million, or $350 million below the red line previously set by the government. Arrears in debt payments to Iraq and Libya and disagreements with Iraq over the price of oil caused interruptions in the flow of oil from Iraq and lengthy delays in the arrival of Libyan shipments. The situation was further compounded by technical problems in the operations of Turkish refineries. In late March 1977 the shortage of oil seriously interfered with power supplies. By that time news of the arrival of an oil tanker in a Turkish port was making front-page headlines in the press.

Under these conditions and facing energy shortages, Turkey had become vulnerable to unwarranted political pressure from Iraq and Libya. The Turks decided to send Çağlayangil to the Gulf in an effort to obtain oil—cheap oil, they hoped—and thus thwart those pressures. Ankara had hoped that oil deals with the Gulf states would show the Iraqis that Turkey had an alternative and would not yield to any pressure. Leaving for the Gulf on 26 March 1977, Çağlayangil was accompanied by the head of the National Petroleum Company. He defined his mission as an effort to diversify oil sources for the long run as a strategy that would also reduce Turkey's dependence on Iraqi oil.

This was not to be. Saudi Prince Fahd promised to see if he could arrange an oil supply under a four- to six-month credit but regretted he was unable to conclude any concrete deal in the absence of Oil Minister Yamani. The Kuwaitis also expressed their willingness in principle to sell Turkey oil under similar credit terms, and then forwarded Turkey's request to expert examiners, where it remained stuck. The United Arab Republic told Çağlayangil that all its oil was spoken for, tied up in long-term export deals, and there was no way the country could make another deal with Turkey. All three nations mentioned in passing the issue of Turkey's relations with Israel.

It was a decisive blow. The Turks had to go back to the Iraqis and agree to all the conditions they set forth.

Kamran İnan, the minister of energy in late 1977, told me that efforts to obtain oil from Saudi Arabia went on even after Çağlayangil's failure, because the difficulties involved in importing oil from Iraq and Libya had not been resolved: "We kept asking them for oil. We tried to use Erbakan again. All their answers were no." İnan had an explanation for the reasons the Saudis refused to provide oil to Turkey at the time:

> The Americans were trying to put the pressure on us because of Cyprus, and President Carter, who has always been hostile to Turkey, was trying to make life difficult for us every which way. Saudi Arabia was then closely linked with the USA. The Saudis were relying on the Americans for protection from domestic and foreign dangers. It was enough for the Americans not to encourage the Saudis to give us oil, for oil not to reach us. They never had to put up any actual pressure. To this, of course, you should add the fact that we had no foreign

exchange at all, which gave the Saudis a very good excuse not to sell us any oil.[21]

Failure in Iran

Turkey's massive reliance on Iraq and Libya, which reached its peak in 1976–1977, was therefore not of its choosing. Not only Saudi Arabia and the Gulf shaikhdom refused to give Turkey preferential treatment; Iran also refused to do so, repeatedly putting obstacles in the way of an oil agreement between the two countries.

Teheran's evasive maneuvers ran contrary to promises made by Iranian representatives at the Organization for Regional Cooperation and Development (RCD) to provide Turkey with preferential treatment on oil exports.[22] The few drops of Iranian oil that did find their way to Turkey in 1974 came through the free market, and the terms were hardly preferential.

The Iranian attitude irked the Turkish press, adding to the anti-Teheran sentiments already prevailing because of tensions that had developed between the two countries for other reasons, as discussed later. In fact, Turkish-Iranian relations were eroding during the early 1970s because of differences on the Kurdish problem,[23] as well as Turkey's tightening ties with Libya and Iraq, whose regimes had no love for the shah's. Even the fact that both Turkey and Iran had a pro-Western orientation at the time as members of both the Baghdad Pact alliance and the RCD was not enough to overcome their mutual jealousy and suspicion, rooted in historical rivalry and ambitions of regional hegemony.

The creation of a new coalition government headed by Süleyman Demirel in April 1975, however, raised hopes for possible change in Turkey's relations with Iran. In early June 1975 Turkey's president, Fahri Korutürk, visited Teheran, and his delegation and their hosts worked out a draft agreement for economic cooperation for the next five years. They also discussed oil imports, and in a press interview on 8 June 1975 the shah spoke about establishing a special committee to examine oil supplies to Turkey. Iranian foreign minister 'Abbas Khal'atbari said Iran was willing to sell Turkey crude oil at OPEC prices under credit terms similar to those Iran gave India.

These statements by the shah and his foreign minister brought no oil to Turkey despite its desperate need. The massive reliance on Arab oil (as distinct from Iranian oil) was becoming a major worry for decisionmakers in Ankara. They felt it was politically important for Turkey to reduce its dependence on the Arabs, particularly with regard to oil. By mid-1975 several Western embassies believed Turkey was trying to use them to exert pressure on Iran to agree to an oil deal.

In late August 1975 another Turkish attempt to reach an oil agreement with Iran failed, and in October that same year the Iranians gave Turkey a final no. On 1 November, during a visit by the shah to Turkey, he stated flatly at a press conference, "Iran is unable at this time to provide Turkey with cheap oil."[24]

There was much aggravation at the Turkish Foreign Office. At the time, Iran was dictating the terms of its economic cooperation with Turkey, and it did so in a forceful, sometimes even an insulting manner. Ankara believed the shah was showing goodwill toward Turkey only in those areas in which he thought he could use Turkey to promote Iranian economic and political interests. Thus he was willing to invest in Turkey's transportation system and food industries because of Iran's supply problems, and he was willing to cooperate in military matters to prevent a Turkish turn toward the USSR and Libya. Oil supply to Turkey, however, was not regarded as a direct Iranian interest, so no action was taken in that area.

In early January 1976 Turkey retaliated by quadrupling transit fees for Iranian trucks crossing its territory. This led to increasing tensions between the two countries, and the Turkish press unleashed another series of volleys against Iran. Again it was claimed that the shah was spending Iran's petrodollars on unneeded weaponry, ignoring Turkey's desperate need for assistance because of the unreasonable increase in oil prices. Iran was also accused of aborting a plan to lay down a gas pipeline from its territory to Europe through Turkey for political reasons and of creating a joint transportation company with Bulgaria to rob Turkish transport workers of their livelihood.

Later that month, while the transit fee crisis had not yet been resolved, the situation deteriorated further. The Turkish press reported, based on information provided by the Enka news agency, that Iranian television stations had accused Turkey of committing atrocities against the Armenians. Turkish television stations retaliated, accusing Iran, among other things, of not allowing citizens of Turkish extraction to use their native language and forcing them to learn Persian instead. Tensions mounted further on 9 February, when Enka announced that an oil deal had been reached between Iran and Greece, purportedly signed on 3 February. The news agency stated that the agreement was signed *after* Iran had turned down Turkish requests with a pretext of previous commitments. Bearing in mind Turkish attitudes toward Greece, this was a particularly bitter pill to swallow.

Curiously, in early 1976, when the international oil market had calmed down, an oil agreement between Turkey and Iran was still not in the cards. Many oil producers were having marketing problems because of the global oversupply, and Iran was becoming more interested in selling oil to Turkey—but not sufficiently so. The oil it did offer Turkey was expensive compared with alternative sources. The shah's tough position thus left Turkey with its uncomfortable reliance on Iraqi and Libyan oil, and its financing problems made it even more vulnerable to political pressure from its suppliers.

In summary, during the years 1973 to 1976 Iran was distinctly reluctant to create a meaningful economic relationship with Turkey, particularly with regard to oil. In an interview I conducted with former foreign minister Haluk Bayülken on 9 May 1985 in Ankara, he said: "The shah has always been jealous of Turkey, but he could never say so in public. He has always maintained that Turkey and Iran were selling the same goods, so they [the Iraqis] could not increase mutual trade. Since we put up pressure, they were feeling a need to do something, so occasionally they were making various offers. The fact that we have been unable to engage them in free trade, since we've always needed special terms, made things all the more difficult."

Financial Difficulties

Troubles with Iraq and Libya

Early signs of Turkey's difficulties in financing oil imports appeared, as mentioned previously, in March and April 1977. In early May, shortly after oil started moving up the Kirkuk-Dörtyol pipeline, Iraq gave first notice that it would be unable to continue delivering oil to Turkey until the latter paid its accumulating debt. During the summer Turkey's financial problems grew worse, and Iraqi pressures mounted. In August 1977, when Turkey's debt neared the $200 million mark, Iraq again warned that it would have to shut down the pipeline. During the fall, Turkey became increasingly concerned about a breakdown in its oil agreement with Iraq.

Although in dire financial straits, during the first eight months of

1977 Turkey increased its imports of crude petroleum by 22 percent, compared with the same period in 1976. Imports of oil distillates also rose significantly; because of the delays in the start of pipeline operations and difficulties financing crude oil imports under the existing agreement, imports almost trebled during the same eight months, again compared with the same period the previous year. This situation put unbearable pressure on Turkey's dwindling foreign exchange reserves. A Turkish suggestion that its oil debt to Iraq be offset against pipeline revenues was promptly turned down.

On 28 August 1977 Energy Minister Kamran İnan went to Baghdad to see if the Iraqi position could be softened. During the visit, which İnan later described as "unfortunate," his hosts took a tough position, refusing to withdraw their threat to shut down the pipeline. Under these circumstances, İnan saw fit to remind the Iraqis of where their water supply originated and noted that the Tigris and Euphrates Rivers would keep on flowing after Iraq had run out of oil. Eventually, the Iraqis agreed to allow Turkey to leave the outstanding debt at $150 million, demanding immediate service of the remaining $54 million.

Following İnan's visit to Baghdad and a meeting between the foreign ministers of the two countries in New York, tensions relaxed somewhat during late September and October, when oil flowed out of Iraq fairly regularly. This changed again in November 1977, however. By that time, Turkey's foreign currency reserves were at their lowest level in six years. The country was unable to keep its debt to Iraq at the agreed $150 million, and the amount outstanding was approaching $250 million. The Iraqis allowed Turkey to run up its debt above the agreed level, but, realizing the problem would only get worse, demanded a payment of $200 million by 16 November. When Turkey was unable to transfer $200 million, as required by the agreements, the Iraqis cut off the flow of oil in the pipeline and with it the loading of Turkish tankers waiting in the port of Yumurtalık. Iraq informed Ankara that oil deliveries would resume once the debt was settled.

Turkey feared a long period of fuel shortages, the first signs of which were already showing in the Istanbul area. İnan said Turkey had reached a point where it was considering weekend bans on motor traffic to save fuel. Soon afterward he announced a decision to use a North Atlantic Treaty Organization (NATO) emergency pipeline to deliver oil to Istanbul from the town of İzmit.

Deputy Prime Minister Necmettin Erbakan, on another pilgrimage in Saudi Arabia, was rushed to Iraq. He met with Saddam Hussein on 27

November 1977 and upon returning to Ankara announced that he had received an explicit promise that Iraq would not keep oil from flowing to Turkey and would not change the preferential terms Turkey had received until then. Thus after twelve days of interruption and a Turkish repayment of $150 million of its debt, Iraq resumed the flow of oil in the Kirkuk-Dörtyol pipeline.

At the same time, similar problems were disrupting the importation of oil from Libya. Turkey's oil debt to Libya reached $50 million in November 1977, and Deputy Prime Minister Alparslan Türkeş was sent to Libya on 27 November to discuss rescheduling. A day before his departure Libya's ambassador in Ankara, Sa'ad al-Din Abu Sharif, gave an inauspicious interview to *Milliyet* in which he accused Turkey of breaking its promise to provide technical assistance to Libya and of leaving other Libyan requests unanswered. As an OPEC member, the ambassador said, Libya was not in a position to sell Turkey cheap oil, but it was able to provide Ankara with project loans and oil purchase credits.

Ankara was outraged by the ambassador's criticism. Energy Minister İnan responded that Turkey owed Libya nothing for its oil. It was paying cash in advance for Libyan shipments, but even then there were frequent interruptions in the delivery of oil from Libya. He flatly denied Libyan allegations that such delays were the result of Turkish debts.

Such protestations did little. A few days after Türkeş's visit the Libyans again refused to load a Turkish tanker, claiming Turkey owed them money. The *Rauf Bei* was held up for nearly three weeks, and loading resumed only after Turkey had paid $9 million to Libya.

The oil supply difficulties were accompanied by growing problems with power supplies. Some of these problems were a direct result of the fuel situation, but the main difficulties were caused by the general shortage of foreign currency, as well as the weather. On 2 November 1977 Bulgaria delivered a note of protest to Ankara, threatening an immediate stoppage of power to Turkey because of a $50 million accumulated debt. An immediate repayment of $10 million was not enough to assuage the Bulgarians. The Ministry of Energy decreed a reduction of television broadcasts to four hours a day, and during the last two weeks in December blackouts occurred throughout Turkey—officially explained by unusually low levels of water behind hydroelectric dams in eastern and central Anatolia. Turkey anxiously began discussions with Syria, offering a release of more Euphrates water in exchange for electricity from the Kapka Dam power station.

Financing problems, fuel shortages, and blackouts forced Turkey to rely on small shipments of oil, financed through loans from foreign companies or short-term borrowing from international banks. For instance, a $135 million loan from the Renault Company was used to purchase oil, and a short-term loan from the Islamic Development Bank covered the importation of 120,000 tons of oil from Pakistan.

On 5 December 1977 *Milliyet* announced that Iraq had agreed to give Turkey more breathing space—namely, a further rescheduling of the oil debt. According to an agreement signed in late November, Turkey was to repay immediately $30 million in cash and another $45 million in export goods and would transfer the rest of the outstanding debt by early April 1978. Following this agreement Iraqi oil deliveries were resumed, but again not for long. Turkey's foreign exchange situation remained basically the same, and a four-month deferment did not constitute a fundamental change. Furthermore, Turkey was unable to abide by the new timetable. Struggling to finance current oil imports (at $5 million a day), the Turks could not repay their debt to Iraq.

On 18 December 1977 an Iraqi delegation came to Ankara to learn which Turkish export goods could be used as repayment in-kind, but it left immediately when it discovered that Turkey had not transferred $30 million as a first cash installment, as agreed in November. Turkey's Central Bank asked for another postponement, to the end of December, but the Iraqis refused. Their delegation, they said, would return to Ankara only after the payment had been made. To emphasize their displeasure, the Iraqis instructed international banks not to forward monies they owed Turkish exporters.

In mid-December the crisis reached new heights. Under a banner headline "Turkey Left with No Oil," Selma Tubel wrote in the *Hürriyet* that Iraq had again stopped the flow of oil in the pipeline, and Libya was refusing to load additional Turkish tankers waiting in the port of Tripoli. The newspaper printed a secret report prepared by the Ministry of Energy warning that Turkey was facing the worst energy crisis in its history. Unless new sources of foreign currency were found, the report warned, chaos would reign. Turkey's stockpiles could cover fifteen days' consumption and no more, and if things remained the same, Turkey could expect a "winter without oil."

Sami Kohen, one of Turkey's most influential columnists, wrote in the Israeli newspaper *Ma'ariv* (27 December 1977) that Turkey was disappointed with the position taken by both Libya and Iraq. Neither country was showing the sympathy Turkey had expected. Kohen also thought Turkey would soon have no fuel unless Libya and Iraq changed their attitude. Shortages of various raw materials were closing down factories, including some very large ones, thus increasing the rate of unemployment and causing social unrest. The political system was restive, too. Twelve members of parliament broke away from the Justice Party, and Demirel's government lost its majority. Under these circumstances discussions with the International Monetary Fund were suspended, and a consortium of foreign banks turned down a Turkish request for a $750 million loan.

Political and Economic Turmoil

Thus at the end of December 1977 Turkey was left at the mercy of Iraq and Libya. The flow of oil from those nations had come to a virtual standstill except for a few shipments, each of which involved lengthy and separate negotiations. On 31 December 1977 Prime Minister Demirel resigned, following a no-confidence vote by parliament. The fuel crisis definitely brought down his government because of its impact on public opinion and, indirectly, its impact on those parliamentarians who defected from the coalition to support the opposition.

On 7 January 1978 a new government was sworn in headed by Bülent Ecevit, who immediately started to address the fuel crisis. With the change of government, a fierce public exchange of recriminations took place. One day after assuming his new position, Minister of Energy Deniz Baykal told the press he had inherited an acute shortage of fuel and energy, despite optimistic statements made by the same office a few days earlier. This irked the outgoing minister, Kamran İnan. He told the *Tercüman* that the outgoing government had left the incoming one with 600,000 tons of crude oil and a fifteen-day stockpile of distillates, which he said should last until the end of January. Furthermore, on the day he left office (16 January), he said, he had signed oil import agreements amounting to 3 million tons.

Baykal retaliated by accusing the former government of "making mistakes in utilizing Turkey's natural resources," which had brought about the energy shortages. Members of the previous coalition evened the score by accusing Ecevit's government of making Turkey "a country of shortages and a nation standing in lines."[25] And indeed, as the new minister of energy was making heart-warming statements about refineries in full swing, people were finding it increasingly difficult to obtain fuel.

All of these recriminations notwithstanding, by the end of January 1978 the oil debt reached a new height—$330 million. The new administration was no more capable of solving this problem than its predecessors had been.

Iraq again stopped the flow of oil to Turkey in early January 1978.[26] The Iraqis even refused to allow the Turks to draw oil from the pipeline storage facilities in Yumurtalık, Turkey, where 350,000 tons of oil (enough for twelve days' consumption) were kept.

On 2 February *Tercüman* drew a bleak picture of Turkey's oil situation. The Iraqi stoppage had brought stockpiles down to one week's consumption. Refineries were operating at less than normal capacity, and Turkey was relying on oil shipments from Iran, Egypt, Syria, and Venezuela, arranged through various brokers. Oil supplies were discussed with Saudi Arabia, Kuwait, and the Soviet Union, but the discussions came to naught because of Turkey's credit problem.

Generally, Turkey's financial situation remained untenable. In early February 1987 debt arrears reached the $1.5 billion mark, of which around 20 percent was oil related. Turkey had to stop all foreign currency payments except for current oil shipments. Its inability to raise the $75 million demanded by Iraq as a condition for the resumption of oil flow in the pipeline made international financial institutions even more concerned, and consequently, raising money became more difficult. On 12 February the leader of the opposition, Süleyman Demirel, was quoted as saying that within a short time the government would be unable to provide citizens' most basic needs.

Libya to the Rescue

Just as the crisis was becoming unmanageable, a marked change for the better took place in Turkey's relations with Libya. In the second half of January 1978 Libya suddenly indicated its willingness to respond favorably to a request by Ecevit's government for a significant extension of Turkey's oil credit line. The "special relationship" with Libya, developed by the new prime minister and his minister of energy back in 1974, was now Turkey's lifesaver.

On 17 and 18 February the Turkish press reported that Libya had started delivering crude oil to Turkey without asking for immediate cash. Two tankers, the *Ata* (55,000 tons) and the *Rauf Bei* (85,000 tons), were promptly loaded in Libyan ports, with payment due only in the spring of 1978. Since then, Libyan oil delivery has been regular.

Reports of Libyan willingness to help Turkey overcome its crisis appeared with increasing frequency as February neared its end. On 22 February Libyan prime minister 'Abd-al-Salaam Jalud visited Turkey. During his stay he agreed in principle to reschedule Turkish oil debts, extend more credit for future oil sales, and provide guarantees for a loan Turkey was trying to obtain in the international capital market. Turkey's request to increase the amount of Libyan oil it received from 3 to 5 million tons, however, was met with a vague promise of willingness for further consideration.

Agreements were then signed that increased Turkey's credit significantly, greatly relieving Turkey's financial problems. Jalud's statements during his visit about Libya's determination to help Turkey overcome both its debt repayment problems and its oil shortages gave Turkey some breathing space. Even though Jalud made it plain that all understandings in principle were subject to approval by his government, it was obvious that Libya was determined to help Turkey.

Libya's generosity was not unreciprocated. On 24 February 1978 *Hürriyet* printed an itemized list of Libya's requests from Turkey, all of which were political and military in nature. Libya, the newspaper claimed, had asked for Turkish support for the "Refusal Front" (as the Arab nations opposing the newly emergent peace process called themselves) in Middle Eastern politics; a change in Turkey's bilateral policies vis-à-vis Israel and Egypt; closer relations with the Palestine Liberation Organization (PLO);[27] support for Libyan initiatives in international forums; and closer military and technical cooperation, including training of Libyan military personnel in Turkey. These demands (further discussed later) did not deter Turkey from entering into the credit agreement. Libya was, it seemed, the only nation in the world willing to help Turkey in its hour of despair, and Turkey was not about to bring to the fore political issues that at the time were low on its agenda.

Toward the end of March 1978 a Libyan delegation went to Ankara to hammer out the final agreement between the two countries according to the guidelines previously agreed to with Jalud. During the discussions Libya indicated that is was unable to respond to Turkey's request to increase the annual supply of oil to 5 million tons because of previous commitments. But the Libyans kept their word on the issue of credit.

On 7 and 8 April an agreement was initialed in Ankara whereby the Libyans provided Turkey with a $450 million credit overall. This included $300 million in oil credits for five years as part of an arrangement whereby Turkey was to pay in cash only 80 percent of any sum due for oil purchases, the remaining 20 percent becoming a Libyan loan; a $100 million "soft" loan to be raised in the international capital market under Libyan guarantees; and a $50 million oil debt rescheduled for another year. It was also agreed that Libya would import $50 million in Turkish goods as partial payment for oil. Official statements spoke of wheat, flour, beef, fruit, and cooking oil, but the press added to this list "strategically significant industrial facilities." The agreement was signed three weeks later in Libya and came into force on 2 June 1978, following an exchange of instruments of ratification between the two Foreign Offices.[28]

Shortly after ratification, on 17 August 1978, Turkey received the $100 million loan for a five-year period at a lower-than-usual interest rate. This was the first external loan Turkey had managed to raise in 1978, and Libya's guarantees clearly secured it for the Turks. The loan also had a significant impact in impressing other financial institutions, which showed more willingness to lend Turkey money during the remainder of the year.

Although Libya's financial assistance brought some relief, it could not save the situation as long as Libya was the only country helping Turkey. By June 1978 Turkey had again reached dire straits. The Iraqi oil supply had not resumed. And the Soviet Union, Iran, and Saudi Arabia refused to enter into an oil agreement with Turkey as long as they were owed money. Thus during the first half of 1978, Libya was Turkey's only source of "contract oil." Everything else had to come either from the spot market or through various brokers, and it was sometimes difficult to make ends meet.

In early July 1978 the *Middle East Economic Digest* reported an "acute" shortage of fuel in the Istanbul area. Traffic on the Bosporus decreased by 16 percent within a week, industry was working at 45 to 60 percent capacity (compared with the second half of 1977), the Ataş refinery was running at 40 percent capacity, and long lines were stretched in front of gas stations. It was clear that Libya's assistance was not enough, and the government had to act swiftly to find other sources of oil to pull the country out of the seemingly endless crisis.

Yet despite the difficulties that reappeared in June 1978, Libya's assistance during the previous months was highly important for Turkey.

The Ecevit administration had inherited huge problems with each of Turkey's oil sources: Iraq, Libya, and the international oil companies. By improving relations with Libya and receiving its support, the problems with one of these sources were solved in terms of both supply and credit. Libya's assistance came during the bleakest period, when a new and relatively inexperienced government—at least as far as oil supply was concerned—was taking its first hesitant steps. For Turkey's energy sector, the Libyan assistance was the only ray of hope during the first half of 1978.

Libya's decision to support Turkey seems to have been a direct result of the unique chemistry developed previously between Libya's leaders and the once and future Turkish leadership. Those special relationships had been formed in 1974, when Libya supported Ecevit on the issue of Cyprus, and were further enhanced by the developing ties between Ecevit's Republican People's Party and Libya's Arab Socialist Union at a time when the Justice Party ruled Turkey. The Libyans apparently believed that with their friend Ecevit in power and the pro-Western Demirel in opposition, political gains would come their way more readily, and they were willing to provide generous credit to further that cause. In view of deteriorating relations between Ankara and Baghdad, the Libyans must have believed that by siding with Turkey and indirectly thwarting Iraq they were gaining "exclusivity" in influencing Turkey's Middle Eastern policies, as further discussed later.

Ecevit's New Policies

Turkey's oil import crisis during 1977–1978 was clearly financial in nature. Between November 1977 and July 1978, there was no physical impediment (such as war, bad weather, or an oil embargo) to a regular flow of oil to Turkey. Further, there was no bilateral political obstacle. On the face of it, Turkey was promised all the oil it needed, including an option to increase the amounts imported from both Iraq and Libya. The only problem, then, lay in financing—but it was a huge one.

The unceasing pressure Turkey tried to exert on Arab oil producers in an effort to improve their credit terms was only partially successful. Iraq twice agreed to postpone oil debt repayments for four to six months. The Libyans took a tough position until early 1978 but were surprisingly generous once Ecevit came to power on 7 January of that year. Saudi Arabia, Kuwait, and Iran were adamant in their refusal to provide oil to Turkey under terms similar to those offered by Libya and

Iraq. Among other things, those oil producers must have feared that meeting Turkish requests would encourage other Third World nations to apply similar pressure.

The dominant theme throughout this period of crisis was frequent interruptions, up to long stretches of complete standstill, in the flow of oil from Iraq. The main lesson the Ecevit government learned from the crisis was that Turkey had to find a way to finance oil imports that did not involve scarce foreign currency. Thus since the summer of 1978 Turkey has relied on a policy of barter, preferring to mortgage future exports of certain commodities against assured energy supplies. This new policy, as discussed in Chapter 4, has created a direct linkage between Turkey's ability to market its export goods and its ability to obtain oil from the producing nations.

Intriguingly—twenty-six years after Turkey's initial military intervention in Cyprus and twenty-two years after the heyday of Muammar Qaddafi–Ecevit relations—these two leaders are still the heads of their respective nations. But now they are far apart. Ecevit's closest political partner in the international arena at one time in recent years was former Israeli prime minister Ehud Barak.

Notes

1. *Pulse,* 10 April 1974.
2. *Pulse,* 8 April 1974.
3. See Figure 2.5.
4. *Milliyet,* 18 February 1975; *Middle East Economic Digest,* 16 September 1977; *Christian Science Monitor,* 25 November 1977.
5. Interview with the author, Ankara, 17 May 1985.
6. See *Briefing,* 23 February 1976.
7. *Tercüman,* 26 April 1976.
8. Interview with the author, Ankara, 17 May 1985.
9. *Pulse,* 25 January 1974.
10. *Pulse,* 25 January 1974.
11. *Günaydin,* 26 January 1974. Saudi political pressure is discussed further later.
12. *Yankı* magazine, April 1974.
13. As "reliable sources" told *Pulse.* See editorial on 26 April 1974.
14. This forum included senior cabinet ministers, military leaders, and top officials.
15. Several items in *Hürriyet,* 2–9 May 1974.
16. *Milliyet,* 5 May 1974.

17. *Milliyet, Cumhuriyet,* 5 May 1974; *Yankı* magazine, 14 May 1974.

18. *Cumhuriyet,* 13 March 1974.

19. Discussion in parliament, 3 May 1974.

20. Presumably, Çağlayangil's personal and family connections in Saudi Arabia played a part in his choice for this mission.

21. Interview with the author, Ankara, 9 May 1985.

22. The RCD members were Turkey, Iran, and Pakistan.

23. Turkey, ever fearful of rising Kurdish national sentiments within its borders, was comfortable with Iraq's anti-Kurdish activities. Iran, on the other hand, was aiding the Kurdish Barazani faction in hopes of weakening the Iraqi Ba'ath regime.

24. Press conference of the Shah, *Pulse,* 2 November 1975.

25. *Pulse,* 9–10 January 1978.

26. This time the stoppage lasted until August 1978. When the flow of Iraqi oil resumed, Baykal admitted that Turkey had received no oil from Iraq since January.

27. During Jalud's visit, Ecevit had declared that Turkey would allow the PLO to open a legation in Ankara, and the Libyan prime minister thanked him profusely.

28. This was just a part of an overall agreement for commercial and financial cooperation. In May 1978 the two nations reached agreements on cooperation in industry, agriculture, tourism, transportation, and technology, as well as the employment of Turkish workers in Libya.

4

BARTERING FOR OIL

Faced with overwhelming problems in their efforts to import oil by conventional means, the Turks decided to try other means. The first indications of the forthcoming change were seen during Süleyman Demirel's premiership (1975–1977). The November 1977 agreement that deferred payment of Turkey's oil debt to Iraq included a Turkish undertaking to cover $45 million of this debt with export goods rather than cash—a marked departure from a long-standing Turkish policy to avoid such deals.

With the 7 January 1978 change of government, the pace of change was accelerated, and the lack of foreign exchange was only one reason for the increased pace. The new prime minister, Bülent Ecevit, believed barter deals could help Turkey find new markets in the oil states, thus helping the development of both industry and agriculture. In this belief, as we shall see, he was fully vindicated just a few years later. In fact, it was part of his overall view of the role Turkey should play in regional development: the country should be a vehicle of regional integration. Ecevit saw the amalgamation of Middle Eastern economies as a precondition for economic growth and political consolidation.

Just a few days after the new government had come into power, the Turkish press began publishing stories about intensive efforts to exchange wheat, fruits, vegetables, and industrial products for Iraqi oil and to provide technical assistance and industrial goods in exchange for Libyan oil. In February 1978 Turkey was given a chance to implement its new policy. During a visit to Ankara, Libyan prime minister 'Abd-al-Salaam Jalud agreed in principle to a proposal by his hosts that Libya would receive Turkish goods in exchange for some of the oil it sold to Turkey. In early April the barter deal was agreed upon: the Libyans agreed to import $50 million in Turkish goods as partial payment for their oil.

The barter policy gained momentum from then on. In the summer of 1978 Turkey signed a series of agreements with the USSR, Iran, and Iraq. Those producers, having realized that Turkey had no other way to purchase the oil it so badly needed, adjusted to the new circumstances and carefully selected those Turkish goods they desired most. The most wanted crop by far was wheat.

Agreement with the USSR

During 1977 the Soviet Union sent Turkey just a few shipments of oil, paid for in foreign currency. In December 1977 the USSR turned down a Turkish request for a bilateral oil agreement.

Once the Ecevit government was in power, efforts directed at the Soviet Union became more intense. Shortly after taking office, Ecevit and his foreign minister, Gündüz Ökçün, met with the Soviet ambassador in Ankara and renewed Turkey's request for an oil agreement. The new leaders made it clear that they were talking about oil imports as part of a more comprehensive agreement, because no oil-for-cash deal would have solved any of Turkey's problems.

The Soviets did not reply immediately, but a few weeks later a trade delegation went to Ankara on a fact-finding mission. Arriving in mid-March 1978, the delegation indicated a willingness in principle to enter into barter deals with Turkey, but their requests were specific. In exchange for oil the Soviets wanted wheat, tungsten ore (an important ingredient in the production of special steels), and other minerals.[1] This was a problem for the Turks. The USSR wanted to secure large quantities of wheat through a long-term (three- to five-year) agreement, but the Turks were unable to commit to the quantities in question because of the uncertainties inherent in agricultural production. Tungsten was a problem, too, because it is regarded as strategic material. The Ministry of Defense and the military objected, as did Turkey's friends in the West.

The agreement, finally concluded on 7 April 1978, called for a 50 percent increase in trade between the two nations but said nothing about either wheat or oil. Both parties realized those commodities required high-level discussions.

Nonetheless, it seems tungsten was mentioned in the agreement. The left-wing *Aydınlık* wrote that the agreement included a clause about exports of 300 tons of tungsten, worth $2 million, to the Soviet Union. Previous trade agreements had mentioned tungsten in passing, but this time Turkey took active steps to realize the deal after a long period in which the Demirel government ignored such references. In early June 1978, when Prime Minister Ecevit was about to leave for Moscow, Energy Minister Daniz Baykal gave an instruction to upgrade the exported ores as required by the Soviets. The Turkish media attributed this move to politics rather than economics. Ecevit's government, unlike Demirel's, seemed determined to improve relations with the USSR in all aspects, and the press saw the decision to upgrade tungsten ores as part of that general approach.

Returning from Moscow in late June, Ecevit was able to announce an important achievement. The Soviets, he said, would provide Turkey with 3 million tons of oil a year starting in 1979 following an agreement reached between the two governments. The negotiations were facilitated by the personal intervention of the prime minister. Naturally, the Turkish press began to speculate about the concessions Ecevit had given Moscow that made this agreement possible.

The senior *Milliyet* commentator, Abdi İpekçi, wrote on 25 June 1978 that Soviet officials only agreed to sell Turkey oil in exchange for wheat and tungsten ores—"nothing else would have done." And indeed, two weeks later Minister of State Hikmet Çetin said Turkey would pay for Soviet oil with wheat and "nonferrous metals." He added, "If we shall be unable to pay the full amount with these commodities, the parties will meet again to decide on more export goods. Should they be unable to close the gap, we shall pay cash."[2]

Other commentators saw this economic breakthrough as a bonus of the political agreement signed by the parties. For the Soviets, the political highlight was a commitment by both parties not to allow foreign powers to establish military bases "of an aggressive or subversive nature" in their territories. In other words, Turkey agreed not to allow the establishment of more U.S. bases, which it was required to do through its 1976 agreement with the United States. Another Soviet achievement was the declaration that the border between the two countries would become "a border of peace and friendly cooperation," which meant it would be open it to all kinds of transportation—something the Soviets had wanted for a long time. Yet another interpretation was that the deal had been made possible because Ecevit had agreed to let the Soviets prospect for oil in Turkish territory.

In reality, the linchpin that made the deal possible was undoubtedly Turkey's agreement to export wheat. The USSR was well aware of Turkey's political instability and must have known that long-term political commitments would not amount to much. The Soviets, for their part, could stop the flow of oil any time Turkey failed to deliver wheat, which represented a warranty of a kind. Without it, it is doubtful that the USSR would have entered the agreement.

Indeed, wheat was Turkey's chief means of payment for Soviet oil. The agreement called for 3 million tons of oil, worth $300 million, to be exported to Turkey. The Turks undertook to export 2 million tons of wheat—more than the entire Turkish wheat export in 1978, a bumper year for crops. The commitment was risky, bearing in mind that Turkey had other wheat export contracts in effect. Furthermore, the Turks were interested in entering into similar barter deals with Iran and Iraq, which were also mainly interested in Turkish wheat.

In summary, Turkey's deal with the Soviet Union had several remarkable features compared with previous oil deals. First, the involvement of the highest politician in the land brought about a dramatic change in the nature and scope of Turkish commitments. Second, this was Turkey's first oil agreement based entirely on barter, with cash serving only as a last resort. Third, in addition to the oil agreement, other agreements of political and strategic significance were signed, and they have had far-reaching consequences for Turkey's relations with the Western powers.

Still, the deal with the USSR (I discuss its implementation later) was unable by itself to solve Turkey's oil supply problems. The Turks urgently needed at least one more oil deal of similar scope to satisfy their needs. They turned their efforts toward Iran, a nation that had always preferred Turkish commodities in exchange for its oil because of its chronic shortages of foodstuffs and other essentials.

An Oil Deal with Iran

Discussions of an oil agreement with Iran became intensive in June 1978, when the Iranians let Turkey know they were willing to exchange

1 million tons of oil for commodities. The timing of this message could not have been better for Turkey. At the time, the oil shortage was being felt, and long lines were stretching in front of gas stations. Even a government threat to nationalize the Ataş refinery failed to persuade the foreign companies operating in Turkey to resume oil imports. Turkey's oil debt reached a new height, more than $500 million.

In early July 1978 Minister of State Çetin went to Iran to enter an oil agreement that would establish mutual trade on a long-term basis. It was agreed that Iran would supply Turkey with 1 million tons of crude petroleum and half a million tons of diesel fuel a year in exchange for cement, glass, wheat, beef, citrus fruit, and other foodstuffs. Formally, Iran would open a $150 million credit line for Turkey, to be covered by those commodities. The terms of the credit were not made public, but the press indicated that it entailed 6 percent interest for fourteen months. On 9 July Radio Teheran confirmed that a barter deal had been signed and expressed hopes that it would contribute to improved relations between the two nations.

Thus Turkey had in place yet another component in the web of barter agreements it was striving to weave around itself. In another sense, this was a significant departure from a long-standing policy that just four years earlier had caused Turkey to turn down an Iranian inquiry about entering into such a barter agreement; at the time, a source in Turkey's Foreign Office had been quoted as saying that such agreements were anathema to the Turks. But under the new realities created by the ongoing financial crisis, this model now best suited Turkish needs.

The scope of the deal was determined by Turkey's industrial and agricultural capabilities in terms of commodities that interested the Iranians—the same factor that has determined all of Turkey's barter deals. Still, the amount of 1.5 million tons failed to satisfy Turkey's fuel needs.

An Oil Deal with Iraq

Following their success with these barter deals with the USSR and Iran, the Turks again turned their efforts toward Iraq. They were hoping to reach an agreement for oil debt repayment in commodities and to then develop a barter agreement for future oil supplies.

While talking with an Iraqi economic delegation invited to Ankara, the Turks were surprised to learn that it had received no mandate to discuss the oil debt; rather, its terms of reference included only the selection of Turkish goods suitable for the Iraqi market, to the tune of just $75–$80 million a year. The Iraqi delegates insisted that the oil debt be discussed separately in Baghdad. The official statement issued following the conclusion of the talks said the Iraqi delegation discussed import agreements on a series of commodities and also made necessary preparations for an intergovernmental agreement on matters pertaining to the oil debt and the resumption of oil supplies to Turkey. The press reported contracts had been agreed for the exportation of foodstuffs, cement, glass, and a number of industrial products (including 40,000 refrigerators). The Iraqis sought to increase the amount of wheat they were importing at the time, whereas the Turks—already overcommitted through their agreements with Libya, the USSR, and Iran—preferred to sell industrial products, particularly electronics.

Iraq preferred to discuss oil at a higher political level. The message was conveyed that a visit by Prime Minister Ecevit, sooner rather than later, would facilitate a solution to the problems dividing the two parties, including the issue of oil. Baghdad apparently sought to place oil discussions in a wider context—encompassing the gamut of issues waiting to be sorted out between Turkey and Iraq—and felt that only at the highest political level could agreement be reached on a package deal that would also include the Kurdish problem, Euphrates water, land transportation, and wheat exports.

Furthermore, the Iraqis were mindful of the agreement Ecevit had signed in Moscow. Suspicious of Turkish promises, they were looking for specific commitments by the prime minister, like those given to the Soviets, before they proceeded.

One reason for Iraqi concern was the scope of Turkish wheat export commitments. The Iraqis recalled that their 1979 request to import 400,000 tons of wheat had been turned down at the technical level: Turkey's Soil Product Office refused to allow more than 130 to 135 thousand tons because of previous commitments and the usual uncertainties involving future crops. Also, indications that Turkey wanted to separate water discussions from oil talks enhanced Iraq's feeling that only direct intervention by the prime minister would sort out the complexities of the entire situation.

Only after the Iraqis had been promised that Ecevit would visit Baghdad was another delegation sent to Ankara, headed by Iraqi minister of planning 'Adnan Hussein Hamadani. Arriving on 22 August 1978, the group was authorized to discuss debt repayments and supply resumption. The Iraqi News Agency announced that same day that Prime Minister Ecevit had accepted an Iraqi invitation to come to Baghdad and had promised to do so soon.[3] Ecevit took part in the discussions with Hamadani, and his personal intervention seems to have made progress possible.

The discussions focused on wheat and water supplies in exchange for oil supplies. Turkey agreed to Iraqi demands regarding the amount of wheat to be exported, and a decision seems to have been made to allow more water to flow out of the Keban Dam, even though no official statement was made to that effect. Energy Minister Baykal was only willing to say that Turkey would "give due consideration" to Iraq's water needs and that a joint committee would soon be set up.

On 25 August 1978 the parties signed a detailed agreement for the flow of oil to resume. Its highlights included (1) a supply of at least 6.2 million tons of oil to Turkey by the end of 1979 (1.2 million tons in 1978 and 5 million in 1979); (2) the entire Turkish oil debt (nearly $300 million) to be paid to Iraq in agricultural and industrial commodities over a period of three and a half years; (3) the Kirkuk-Dörtyol pipeline to resume operations at full capacity, with Iraq paying Turkey passage fees for 15 million tons even if for some reason full flow became impossible; (4) future payments for oil to be purchased by Turkey to be in cash, with a credit line of up to $50 million; and (5) Turkey to export to Iraq 1 million tons of wheat over the next three years if crops remained sufficiently large.

This was a major breakthrough in relations between Turkey and Iraq, as well as a solution (at least over the short term) to Turkey's oil supply problems. Once the agreement was in place, Turkey had secured 12.5 million tons of "contract oil" for 1979—from Libya, the USSR, Iran, and Iraq. That quantity, added to domestic production, should have been sufficient for Turkey's needs. Baykal declared that the Iraq agreement would close a major gap in Turkey's oil procurement plans, whereas the exports earmarked for debt repayment would contribute significantly to industrial growth and stabilize the balance of trade.

Faced with an ominous foreign currency reserve situation and mindful of the huge quantities of wheat already promised to other oil producers, the Turks had to accede to Iraqi demands to discuss oil in a much wider context—namely, the entire scope of bilateral relations. Thus Turkey's water commitments put in danger development plans in the southeastern part of the country, while the agreement with Iraq on the Kurdish problem could have had a deleterious effect on Turkey's relations with Iran. Also, the amounts of wheat already promised to others made it impossible for Turkey to enter into a direct barter agreement with Iraq, as it preferred. The Turks were forced to promise cash payments for future oil deliveries, knowing full well how difficult making such payments would be.

The Politics of Oil and Economics of Wheat

Against a background of unrelenting difficulties in finding and financing sources of oil imports during the first half of 1978, the oil agreements with the USSR, Iran, and Iraq must be regarded as a diplomatic triumph. Turkey had managed to secure quantities of contract oil sufficient to meet its needs while committing the smallest possible amount of foreign currency in payment. But those agreements had yet to pass the acid test of implementation. Turkey had undertaken enormous obligations, and it was not clear whether it could meet them, as dependent as doing so was on future agricultural crops.

The four oil agreements signed in the spring and summer of 1978 with Libya, the USSR, Iran, and Iraq together called for an annual export of 3 million tons of wheat starting in 1979. In 1978, a bumper year for Turkish wheat, 1.9 million tons were exported to fifteen countries; Libya was the only oil producer to receive a significant share in that export. Now such countries as Italy, Tunisia, Pakistan, Poland, Romania, and Bulgaria—major destinations for Turkish wheat in 1978—had to be eliminated from the list, even though negotiations with some of them were almost concluded. The need to import oil and foreign exchange constraints were dictating Turkish export policies, and Turkish foreign policies had also become subjugated to oil.

The 1979 wheat export commitments, undertaken against expert advice, were a direct consequence of political intervention at the highest level in concluding the 1978 oil agreements. The change in export destinations was largely a political problem. Prime Minister Ecevit wanted to resume having a regular supply of oil, and only he could twist the arms of Turkey's technical experts on the issue of wheat commitments. Founded as they were on (vain) hopes for unusually large crops, the wheat agreements had a potentially major effect on Turkish foreign policy.

The series of agreements Ecevit signed in 1978 constituted an abandonment of two principles that until then had guided Turkey in all its oil negotiations: (1) to avoid including political or quasi-political issues (such as the Kurdish situation or water problems) in oil discussions and (2) to keep strategic issues (such as military bases, transfer of knowledge, and sales of strategic materials and equipment) away from such discussions. At the cost of relinquishing these principles, Turkey had acquired some breathing space: Iraqi oil supply was resumed, the oil debt due for immediate repayment (between August and November 1978) was cut in half, and the foreign currency reserve situation was improved. Yet this breathing space was limited. The global conditions that made the agreements possible in the first place—an oversupply of oil, a temporary weakness of the U.S. dollar, and a worldwide shortage of wheat[4]—soon changed, and with them changed both the oil producers' motivation to abide by their agreements and Turkey's ability to fulfill its obligations. By the fall of 1978 oil shortages again loomed large on the horizon. The turmoil that took place in Iran in November 1978 and the global oil crisis that followed made Turkey's position much weaker and jeopardized its 1978 oil agreements.

Indeed, the international scene in 1979 and 1980 was overshadowed by two successive energy crises, brought about by the Iranian revolution and the outbreak of the Iran-Iraq War. The impact of these crises on Turkey is examined in detail later, but it should be noted here that they put Turkey's oil agreements to a severe test.

Ecevit and other members of his coalition took great pride in their ability to secure enough oil for Turkey for 1979 and 1980, despite the grievous economic situation. At a press conference held shortly after he announced retirement on 14 November 1979, Ecevit explained that Turkey's ability to win contracts for the supply of its entire oil requirement would "relieve it of [a] dangerous dependence on expensive, free-market oil." The incoming prime minister, Süleyman Demirel, saw things differently. As he told *Milliyet* on 16 November 1979, "The meaning of this contractual obligation is that we can get the oil if we can find the money. In other words, Ecevit has found one horseshoe. All that's left for me now is to find three more shoes and a horse to ride."

The idea of changing the main means of payment for oil from foreign exchange to export goods encountered serious difficulties in 1979 and 1980. Domestic problems—such as weather, transportation, harbor services, hoarding, and the like—suddenly became major factors affecting Turkey's ability to import oil. First and foremost among these difficulties was the ability to export wheat, the currency on which Turkey's leaders had pinned their oil financing hopes.

Implementation of the Soviet Agreement

Negotiations over the implementation of Turkey's economic agreement with the USSR were facing serious obstacles at the start of 1979, even before the flow of Soviet oil to Turkey was to begin. First, Turkey had backed down from its obligation to export tungsten ores to the USSR. Under Western pressure Turkey broke the barter deal and sold its ores to West Germany instead, where they were used in the production of Leopard tanks. Moscow was furious, and talks proceeded in a chilly atmosphere from then on.

But the major obstacle to implementation of the barter deal was wheat. Aside from disputes over the price of wheat or the varieties the Soviet Union preferred, the main issue was Turkey's inability to supply the quantities promised.

A trade protocol signed by Turkey and the USSR on 30 January 1979 showed an erosion of the original agreement; its scope was narrower, and its objectives had been altered. Originally, according to a reliable source,[5] the parties had agreed to an exchange of 1.5 million tons of wheat for an equal amount of oil over three years, during which time Turkey would be able to purchase another 100,000 tons of oil for cash. The first transaction under this agreement, however, took place only a year later, in May 1979, whereby 40,000 tons of Turkish wheat were exported in return for an import of 50,000 tons of oil.[6]

In all, oil imports from the Soviet Union in 1979 amounted to around 1.5 million tons, about half the quantity Turkey could have purchased had it been possible for it to finance such imports. Of this amount, only 50,000 tons were covered by wheat exports in a direct barter deal. The implication is that most Soviet oil was paid for in cash. The implementation of the original pretentious agreement thus left a lot to be desired.

Bartering with Libya

Political misgivings of the kind that hampered implementation of the barter deal with the USSR were not present in the agreement with the

Muslim producers. Iraq, Iran, and Libya were "sister nations," and no political problem was involved in tightening Turkey's relations with them. Thus the implementation of the agreements with these countries became more successful each year, culminating in 1981 and 1982.

The oil agreement signed with Libya in April 1978 was based largely on cash payments; only 20 percent of Turkey's imports were to be financed by export goods. Between 1979 and 1981, however, the barter component was gradually increased, as more and more Turkish goods were accepted in return for Libyan oil.

Like the other agreements, the Libyan one too faced major obstacles in its initial stages. The revolution in Iran caused a global oil crisis: oil prices were increasing again, and producers were making every effort to get out of earlier contractual obligations. With spot market prices at double the posted cost, Libya could not resist the temptation. Thus in 1979 it supplied Turkey with one half of the contractually promised amount, citing force majeure as the reason. Sworn friendship was set aside.

Turkey suffered major financial damage as a result. It was forced to purchase the balance of its fuel needs, in the form of distillates, on the free market, paying much more than the sums officially agreed to with Libya. The Turks also were unable to make use of the Libyan credit promised them in the 1978 agreement for a larger amount of imported oil.

Yet by January 1980 the energy crisis had ended, and Libya and Turkey entered into a new agreement whereby Turkey would export to Libya 300,000 tons of wheat, 100,000 tons of barley, and a similar quantity of flour during 1980.[7] Turkish officials added that similar amounts would be shipped to Libya in 1981 as well. Undertaking such a commitment to Libya while refusing Soviet requests for grain exports indicated the importance the Turks attached to barter trade with Libya. At the same time, the Libyans were showing more enthusiasm for such deals with the Turks.

Indeed, in 1980 Libya increased its share of Turkish wheat. In 1979 the Libyans had received 18 percent of total Turkish wheat exports, and in 1980 they got more than a third, the largest share among Turkey's trading partners. At the same time, Libya increased the amount of oil it sold Turkey to 2.6 million tons (compared with 1.7 million in 1979). This time the Turks received nearly all the oil agreed upon by the two countries.

During 1981 the volume of barter trade increased dramatically between Turkey and Libya. In February 1981 an agreement was signed whereby Turkish oil debts outstanding as of 31 December 1980 would be repaid over two years through commodities and foreign currency payments owed to Turkish laborers in Libya. It was agreed that 30 percent of the workers' salaries would be transferred to a special account to be used for debt repayment; the Turkish government undertook to repay those deductions to the workers' families in Turkish currency.

Yet another agreement was signed on 24 March 1981, calling for all of Turkish oil imports to be repaid in-kind. Thus the 1978 oil agreement between these two countries was converted into a full-fledged barter deal, which significantly boosted Turkish exports to Libya in 1981. From $43.3 million in 1979 and $60.3 million in 1980, Turkey's exports to Libya rose to $441.5 million in 1981. Still, Turkish commodities were able to repay only 56 percent of imports from Libya that year, which included mostly oil.

Dealing with the New Iran

During the first half of 1979, implementation of the Turkey-Iran oil deal was directly affected by the turmoil in Iran, which brought oil exports to a near standstill. Even though Turkey desperately needed oil and Iran was experiencing large-scale food shortages, both events in Iran and Turkey's marketing difficulties made implementation of the barter agreement between the two countries impossible.

The suspension of Iranian oil exports to Turkey immediately halted Turkish commodity exports to Iran, because the goods most needed for the Iranians—such as wheat and beef—were earmarked for oil payments and consequently were diverted to other oil producers who could deliver. The barter agreement was kept on ice for more than eight months.

Ruholla Khomeini's regime did not cancel the agreement signed with the shah on July 1978, but under the new circumstances its terms had to be revised. In early June 1979 the two parties met for discussions in both Teheran and Ankara. Expressions of sympathy by leaders on both sides and a favorable attitude toward Iran by the Turkish press were early indications of a turn for the better in relations between the two countries. Turkish officials spoke of a "constructive atmosphere" in the talks with their Iranian counterparts, and the press indicated that a "political gesture" by Iran was forthcoming. And indeed, the new oil agreement called for the purchase of 1.1 million tons of oil during the remaining months of 1979, but the crowning achievement was an Iranian commitment to provide Turkey with 5 million tons of crude oil in 1980.

Iran's new trade policy emphasized the need to purchase everything, from shoestrings to industrial equipment, from Muslim countries as mandated by the ayatollah. Therefore Iran turned to Turkey to provide its needs in wheat and barley, leaving Western suppliers on the back burner.

Difficulties with the delivery of Turkish export goods in 1979 and exigencies of the global energy crisis notwithstanding, the two countries were clearly interested in implementing their barter deal. The Turks realized that to obtain a significant part of the oil Iran was willing to sell them in 1980, they needed to give the Iranians higher priority—namely, allocate a larger share of the export goods they asked for, particularly wheat and cement. Iran, for its part, realized that to increase the quantities of foodstuffs imported from Turkey, it should deliver more oil.

A curious phenomenon, highly indicative of the mutual need to barter, was developing along the Iran-Turkey border during 1979 and 1980. Under the auspices of Turkish provincial governors—who gave proper instructions to customs officials—local residents began exchanging produce, processed foodstuffs, and cigarettes for Iranian fuel products. Starting on a small scale in early 1979, this initiative soon solidified into an exchange of many thousands of tons of goods each month, to the extent that it brought about a significant relief of the fuel shortages in southeast Turkey. Unable fully to realize their barter agreement at the national level, the two governments allowed a transfer of some surplus commodities to the border area, and the free market did the rest.

When Iranian "students" took over the U.S. Embassy in Teheran in November 1979, Turkey was perhaps the only nation to reap immediate benefits from the deterioration in U.S.-Iranian relations. The Americans demanded that Turkey participate in the economic sanctions they imposed on Iran, but at the same time the Iranians were making every effort to secure Turkish neutrality in the conflict. Those efforts included offering much better terms for oil purchases.

The stronger U.S. pressure grew, the more generosity Iran displayed, as the international oil market was relaxing again. On 16 April 1980, five days after the Americans had asked Turkey to join the sanctions, three oil tankers left Iran en route to Turkey—even before the Turks had paid for their $95 million in cargo.

On 11 June 1980 the Iranian foreign minister, on a visit to Istanbul, informed his Turkish counterpart that Iran had decided to increase its credit line to Turkey to $500 million, $300 million of which was to be used for oil purchases. It was agreed that the entire loan would be paid with foodstuffs and industrial and agricultural equipment. No other oil producer had given Turkey such comfortable terms for oil imports.

The 1980 trade figures show a breakthrough in Turkey-Iran economic relations. The volume of Turkish exports to Iran increased by a factor of seven compared with the slump the previous year. In 1980 the Iranians received 15 percent of total Turkish wheat exports and 40 percent of barley exports, as well as large quantities of cement and tires; in return, they released 2.6 million tons of oil to Turkey.

In late 1980 the Iran-Iraq War broke out, and again Turkey reaped some windfall gains. The two combatants greatly needed foodstuffs and other commodities, and the list of Turkish exports to both countries was expanded to include goods not previously needed. As early as 3 October *Milliyet* reported that Iran had expressed willingness to sell Turkey as much oil as it required without demanding cash, provided it received more beef and other foodstuffs. On 12 December the Turkish press reported a new barter deal consisting of 400,000 tons of wheat in exchange for 560,000 tons of oil. Even though their oil exports came to a near standstill because of the war, the Iranians made special efforts to maintain oil deliveries to Turkey.

Eager to have Turkey remain neutral in the Iran-Iraq War, the Iranians informed Turkey in April 1981 of their willingness to defer payment for two-thirds of the amount of oil Turkey purchased that year, thus extending the Turks' credit line by another $200 million. It was also agreed that this extra credit would be repaid with Turkish export goods.

The trade statistics for 1981 and 1982 seem to indicate that the Turkey-Iran barter deals were successful. In 1981 Turkey sent Iran 56 percent of its total wheat exports and significantly increased the quantities of beef, flour, tires, and cement it shipped out. Whereas in 1980 Turkey was able to finance only 15 percent of its imports from Iran with goods, in 1981 Turkish exports covered 40 percent of oil imports from that country, and 1982 saw a triumph of sorts: total Turkish exports to Iran were larger than total imports from that country.

The main factors accounting for the expansion of Turkish-Iranian trade were Iran's conflict with the United States and its war with Iraq, both of which made Iran's economic position rather untenable. Under these circumstances Iran had little choice but to allow Turkey to finance its oil purchases entirely through barter and credit. Still, the barter agreement signed between the two countries in July 1978, prior to these political developments, was the springboard for the great expansion of Turkish exports to Iran. Thus Turkey made a major step toward solving its chronic problem of oil import financing.

Dealing with Iraq Again

The August 1978 Turkey-Iraq oil agreement promised the delivery of 5 million tons of oil to Turkey in 1979; in return, Turkey was to deliver to Iraq 400,000 tons of wheat by December 1979, as well as many other industrial and agricultural products. The flow of oil in the Kirkuk-Dörtyol pipeline was resumed, and the Iraqis allowed Turkey to draw the amounts it required.

In the spring of 1979 Turkey again ran into financing problems, to the extent that it was unable to secure for the agricultural sector the fuel needed to harvest wheat. The Turks asked Iraq for help, and in April 1979 the Iraqis responded favorably, giving Turkey a special $40 million credit for the purchase of 337,000 tons of oil. This six-week credit was specifically earmarked for saving Turkey's summer crops.

In gratitude, Turkey's leaders moved Iraq to the top of the list of nations entitled to purchase Turkish wheat in 1979. Around 30 percent of total wheat exports—200,000 tons—was delivered to Iraq that year. Although the Turks were unable to fulfill other export commitments to Iraq, that did not interfere with the regular supply of Iraqi oil. In 1979 Iraq again was the top Turkish oil supplier, having relinquished that position to Iran the year before (see Figure 2.5).

During 1980 the volume of Turkish exports to Iraq did not increase significantly. Even though the Iraqis wanted to import foodstuffs, textiles, and cement, as well as larger quantities of wheat, the Turks were unable to accede to those requests.

The Iran-Iraq War marked a significant turning point in the development of Turkish exports to Iraq. Prior to the war the Iraqi demand for Turkish goods had been limited to foodstuffs and a small number of industrial products, but in October 1980 the list was greatly expanded. The Iraqis asked for tires, metals, vehicles, blankets, and medical equipment and renewed their request for more foodstuffs. In early November 1980 Iraq made a rather unusual request—it asked Turkey for oil distillates. The Iraqi minister of mining and oil went to Ankara to explain that Iraq was suffering severe shortages of gasoline and diesel fuel, so badly needed by its military, because its refineries had been damaged by the war. The Iraqis also asked Turkey to allow shipments of those same products from other countries to cross its territory. In exchange, the Iraqis promised to arrange deliveries of Saudi and Kuwaiti crude oil; they were unable to keep up their own deliveries because the pipeline had been hit.

Thus Turkish exports to Iraq increased by a factor of four within a year. Cement, gasoline, cotton, beef, eggs, buses, and blankets—some in large quantities—joined traditional exports. These exports covered more than a third of Turkey's imports from Iraq, compared with only 11 percent in 1980. As Table 4.1 shows, the great expansion of Turkey's exports to Iraq, Iran, and Libya raised the scope of Turkish exports to Muslim nations by a factor of three in one year.

The consideration both Iraq and Iran gave Turkey was politically and strategically motivated. As tensions mounted between the two countries, Turkey became increasingly important—logistically as an important route for goods and equipment and economically as a supplier of goods needed by both. Neither Iran nor Iraq could afford to antagonize Turkey. Even regarding the Kurdish question, long a major bone of contention between these countries, both Iran and Iraq had to avoid

Table 4.1   Turkish Exports to Muslim Countries, 1980–1981

	1980		1981	
	U.S.$ (millions)	% of Total Exports	U.S.$ (millions)	% of Total Exports
Iraq	134.8	4.6	559.0	11.9
Libya	60.3	2.1	441.5	9.4
Iran	84.8	2.9	233.7	5.0
Saudi Arabia	43.7	1.5	187.4	4.0
Syria	102.9	3.5	129.4	2.8
Jordan	48.1	1.7	97.1	2.0
Lebanon	72.9	2.5	85.5	1.8
Egypt	20.3	0.7	72.1	1.5
Kuwait	50.1	1.7	71.0	1.5
Total	623.0	21.4	1,886.3	40.1

Source: Turkey Institute of Statistics, *Statistical Yearbook of Turkey, 1983* (Ankara: Government of Turkey, 1983) p. 357.

raising Turkish ire. Furthermore, the Iranians had an even more urgent motive: they feared Turkey might join the U.S.-sponsored embargo against them.

Difficulties with the Saudis

Turkey persisted in its efforts to reach an oil agreement with Saudi Arabia, despite the failure of such efforts during the period 1974–1977. As the Organization of Petroleum Exporting Countries' (OPEC) largest and most important exporter and a pro-Western nation that was also devoutly Muslim (and therefore acceptable to all political sectors in Turkey), Saudi Arabia was an ideal partner for Turkey in an oil agreement. Furthermore, since early 1979 the price of Saudi crude oil had been much lower than the prices charged by other OPEC exporters. Nevertheless, the Turks were unable to achieve an agreement in 1979–1980 as well.

The factors that accounted for these failures mainly involved Saudi economic and commercial interests. When Turkish leaders spoke euphemistically of an "incompatibility" between Saudi oil marketing policies and Turkey's bureaucratic rules for oil procurement, they actually meant that selling oil to Turkey was uneconomical for the Saudis in view of Turkey's need for credit or its desire to pay for the oil it purchased in-kind. Thus Turkish offers were unattractive to the Saudis. Once tabled, discussions were suspended in effect throughout 1978, when Turkey was busy working out long-term barter deals with other producers. In late 1978 Saudi Arabia increased its oil production by 25 percent to offset the shortage emerging in the world market, but even then Turkey made no real effort to gain access to Saudi oil.

Such efforts resumed in the spring of 1979 and consisted of a series of visits by Turkish officials to Riyadh. On the whole, these visits constituted an almost exact repetition of the series of failures experienced by Lütfi Doğan, Necmettin Erbakan, and İhsan Sabri Çağlayangil in previous years. The Saudis took pains to avoid actual negotiations. They preferred to use middlemen, spreading rumors and making idle promises, but in the end they did not agree to sell oil to Turkey for various reasons.

First, under the prevailing conditions of the global oil market in 1979, Turkey had no chance to compete with other Western buyers. The Turks kept asking for special (preferential) terms for their deals, while rich Western nations were willing to pay higher-than-posted prices for Saudi oil in cash, which was still better than the cutthroat buyers' competition in the spot market.

Second, Saudi Arabia adhered to OPEC's resolutions requiring that oil be sold only for hard currency because of its general interest in maintaining the strong global position it had attained over the past few years. Saudi Arabia was wary of a wave of requests for special consideration if a precedent was set with a Saudi credit or barter deal with any country. Also, at the time the Saudis were hardly interested in bilateral oil agreements, preferring to market their oil through multinational corporations or other third parties.

Finally, until 1981 the Saudis were not interested in Turkish export goods. They complained of their high prices as well as their low quality.

Still, along with intergovernmental contacts, oil negotiations were conducted through Saudi brokers. Such unofficial discussions, which had begun as early as Ecevit's days, muddied the water further. The Turks were content to work in both channels at the same time, hoping that one way or the other they would be able to obtain 5 million tons of oil a year. The Saudis, for their part, used the split channels as a means to avoid a deal if it turned out to be unprofitable or to receive ample brokerage fees if it went through. This negotiating tactic was common among the Saudis in 1979 and 1980, when the price gap between Saudi Arabia's and other producers' oil—to say nothing of the spot market—was fairly large. Still, everything fell through in the end.

Wheat for Oil—The Turkish Experience

During the years 1979–1981, wheat was Turkey's most attractive export commodity, and a number of nations were clamoring for it. Besides the four oil producers that entered into barter deals with Turkey in 1978, a veritable flood of requests came from various other nations such as China, Italy, Tunisia, and Syria—not all of which were in a position to offer oil in exchange. Some of the nations involved, including the USSR, Iran, and Iraq, made plain their willingness to buy any amount of wheat Turkey was willing to sell. Under such pressure and faced with disappointing crops, Turkey had to be very careful in selecting its wheat export destinations in 1979–1980.

Until 1978, most Turkish wheat had been exported to a number of nations in eastern and western Europe, whereas in 1979 more wheat was diverted to Iraq and Libya. This trend became more significant in 1980 and 1981. Turkey had to extricate itself from previous obligations to its traditional wheat buyers in Europe and North Africa and divert around 80 percent of its 1980 wheat exports and 93 percent of its 1981 exports, to Iraq, Libya, and Iran (see Table 4.2). These Muslim oil producers captured Turkey's wheat surpluses through barter deals because of Turkey's need for oil.

All in all, however, Turkish wheat exports in the years 1979–1981 can only be regarded as a devastating failure, not least in terms of their impact on Turkey's energy problem. In fact, Turkey's inability to meet its wheat export commitments under the 1978 agreements actually worsened its oil import problems. High hopes were pinned at the time on Turkey's success at exporting nearly 2 million tons of wheat in 1978 (of a total crop of 16.8 million tons). The press reported that Turkish officialdom was expecting to repeat this success, if not exceed it, in the period 1979–1981.[8] Official estimates of Turkey's wheat export capabilities ran as high as 3 to 4 million tons in both 1980 and 1981, yielding an expected $500 million a year—"two months' worth of oil imports," according to Prime Minister Demirel.[9]

Turkey's bureaucratic realities, however, made it impossible for the country to export most of its wheat surpluses. A series of domestic problems was responsible for this situation, despite the satisfactory volumes of crops during those years. For instance, most surpluses were owned by the private sector, and the government had a difficult time acquiring them for export. By and large, surplus crops were hoarded, and the rest were smuggled out of the country by individual farmers.

Table 4.2   Turkish Wheat Exports and Share of Muslim Oil Producers, 1978–1981

	1978	1979	1980	1981
Total crops (in millions of tons)	16.8	17.6	16.6	17.0
Amount exported (in millions of tons)	1.9	0.69	0.34	0.32
Wheat export revenues (in millions of U.S. dollars)	208.3	86.2	52.0	53.8
Revenues from exports to Libya, Iraq, and Iran (in millions of U.S. dollars)	57.7	40.9	41.7	49.9
Exports to Libya, Iraq, and Iran as % of revenues	28	47	80	93

Sources: Food & Agriculture Organization (FAO), *Production Yearbook* (Rome: 1980–1982); FAO, *Trade Yearbook* (Rome: 1980–1982); "Turkey's Economic Potential in the Islamic World," *Near East Briefing* (Ankara: 1982), pp. 14–15; Turkish State Institute of Statistics, *Annual Foreign Trade Statistics* (Ankara: Government of Turkey, 1978).

Also, harvesting the wheat and, more particularly, transporting it to harbors required huge quantities of fuel, which Turkey did not have. Finally, Turkish ports were constantly congested; in the absence of proper storage facilities, exportation was at times physically impossible.

Thus the Turks were unable to repeat or even approach their 1978 success. In 1978, wheat paid for 15 percent of Turkey's total oil imports; the figure for 1979 was 5 percent and for 1980 and 1981, 1.4 percent each year.

Nevertheless, the failure of wheat exports did boost the development of other export commodities, both agricultural and industrial, as substitutes in fulfilling Turkey's barter obligations with the oil producers. Fortunately for Turkey, the need to find substitutes for wheat arose at a time when both Iran and Iraq, because of the war between them, were developing a huge demand for such goods.

Direct Oil Import Deals

Any discussion of the bilateral relationship between oil importers and exporters is incomplete without reference to the significant increase in the number of direct oil deals (that is, deals not involving any of the multinational oil corporations or other brokers) in global oil trade. Turkey was one of the nations that made direct oil deals a cornerstone of their international relations during the period discussed here, so I now examine the impact of those deals on Turkey's relations with the oil producers.

Nearly nonexistent in previous decades, direct oil contracts became a worldwide phenomenon during the 1970s. In 1973 direct deals (government to government or to a state-owned oil company) constituted only 7.9 percent of total oil trade; in 1980 (which turned out to be the peak) such deals encompassed nearly 45 percent of the total. This was a result in part of the increase in oil-producing nations' control over their oil assets and their consequent desire to control export destinations and reap for themselves economic and political benefits. Also, at the same time many importers were clamoring for preferential treatment in the face of increasing oil prices, and they were feeling that a direct deal with a sovereign government would make supply more reliable.

As Table 4.3 shows, Turkey had been importing an unusually high proportion of its oil through direct deals. This fact involved certain

Table 4.3   Scope of Direct Oil Purchases, Selected Countries, 1979

Country	Oil Imported Through Direct Deals (million tons)	Total Imported Oil (million tons)	Direct Oil as % of Total Imports
Turkey	9.0	12.6	71
Brazil	30.0	49.6	60
India	11.0	19.2	57
Spain	25.0	49.9	50
France	50.0	135.9	37
Japan	85.0	263.7	32
Italy	22.5	124.6	18

Sources: Ian M. Torrens, *Changing Structures in the World Oil Market* (Paris: Atlantic Institute for International Affairs, 1980); United Nations, *Energy Statistics Yearbook* (New York: UN, 1980–1982); *Petroleum Intelligence Weekly, Platt's Oilgram,* and *Petroleum Economist* (1980).

constraints for the Turkish government, which are discussed later. Briefly, the intensification of direct contacts among the highest political levels exposed Turkey to political and other pressures by the oil producers. In the past, the multinational oil corporations had acted as a sort of buffer that allowed importers to ignore the political agendas of the producer governments. Now things had changed. Also, Turkey was busy solving its short-term problems (securing tomorrow morning's oil supply), whereas the exporting governments could afford a longer view of their own interests; that discrepancy tilted the balance decisively in favor of the oil producers.

Conclusion

The four oil agreements signed during 1978 secured for Turkey a total supply of 12.5 million tons for 1979. Because of Turkey's inability to meet its export commitments, as well as difficulties caused by the global energy crisis that year, fewer than 8 million tons were actually imported through these deals. Turkey had to close the gap by importing large quantities of distillates on the spot market, paying much more than it had paid for contract oil the previous year. This imposed a heavy burden on the Turkish economy, holding back its recovery from previous tribulations.

The new policy that gave preference to payment in-kind rather than in cash for imported oil therefore did not significantly relieve the oil financing problem during 1979. Most oil imports were paid for in foreign currency, which was not readily available, and severe fuel shortages again developed. But despite these financial difficulties, Turkish exports to Middle Eastern oil producers greatly expanded. The failure to export wheat was offset by other commodities required by those producers, especially after the war between Iran and Iraq broke out. Those exports financed an increasing share of Turkish oil imports and at the same time laid a solid foundation for large-scale trading with the Muslim nations of the Middle East.

In the longer run, the patterns established in trade relations with Libya, Iran, and Iraq were also reflected in relations with Saudi Arabia. Sales of Saudi oil to Turkey resumed in 1981, and a counterflow of Turkish exports to that country was growing. Turkish exports to Saudi Arabia increased fourfold during 1981, amounting to $187.5 million, enough to finance 46 percent of total Saudi exports to Turkey. In 1982 the volume of Turkish exports to Saudi Arabia doubled again, reaching $358 million and financing 71 percent of imports. A similar, although more moderate, development took place in Turkey's trade with Kuwait.

Thus in the early 1980s the policy of engaging in barter trade was vindicated. It forced Turkey into an accelerated development of various industries that in the end brought about Turkey's greatest economic achievement since the outbreak of the energy crisis—a sizable expansion of exports. Politically, the decision to give Muslim oil producers preference over others as destinations for Turkish exports, although imposed by the exigencies of the energy crisis, gave more substance to Turkey's relations with those countries. In many respects this was an important political achievement, because that rapprochement barely affected Turkey's relations with the rest of the world.

Notes

1. On 17 March 1978 *Hürriyet* wrote that the USSR was interested in uranium as well, and other newspapers spoke about high-grade iron ore.
2. Quoted in *Pulse,* 11 July 1978.
3. Ecevit did not reach Baghdad until 2 December 1978.
4. The causes of the world shortage of wheat were particularly bad crops among the larger producers and China's entry into the world market as a gigantic importer, to the tune of 10 million tons a year.

5. *Middle East Economic Digest (MEED)*, 12 October 1979.

6. *MEED*, 18 May 1979. These numbers were subsequently confirmed by Minister of Trade Teoman Köprülüler.

7. Data for grain exports in this chapter from Ministries of Trade and Agriculture.

8. As an indication of Turkey's unrealistic expectations, Minister of Agriculture Cemal Külahlı predicted (in an interview with *Tercüman*, 6 February 1981) that wheat would cover 50 percent of Turkey's total oil imports.

9. *Hürriyet*, 24 May 1980.

5

INTERNATIONAL CONSEQUENCES OF THE 1980 CRISIS

Despite the relative success Turkey achieved in its barter oil deals, the social-economic crisis that had been plaguing the country since the mid-1970s was unrelenting. Among other things, the crisis was a chief cause of the military coup that took place in Turkey in September 1980. Before moving on to discuss Turkey in the 1980s, I examine 1980 itself, a seminal year in many respects. In particular, the oil crisis that hit Turkey (and the rest of the world) during that year and the outbreak of the Iran-Iraq War have had far-reaching consequences.

The Global Energy Crises

For the purpose of this discussion, the terms *global energy crisis* and *energy problem* must be rigorously defined. In a global energy crisis most oil importers feel constrained, individually or as a group, to introduce major policy changes to adjust to a sudden change or the threat of such a change in the level of oil supplies, in oil prices, or both.[1] An energy problem is defined as a difficulty an individual importer encounters in providing for its energy needs at the national level. Such a problem may therefore arise independently of any fluctuations in the global energy sector.

Since 1973, three global energy crises have taken place:

- An oil embargo crisis began on 16 October 1973, when many oil exporters imposed an oil embargo on the United States and the Netherlands, allegedly for supporting Israel in the Yom Kippur War that was raging at the time. This crisis ended in spring of 1974, when most of the exporters called off the embargo.
- The Iranian revolution brought about a crisis that began on 22 October 1978, when Iranian production was first disrupted. The crisis lasted until the end of 1979.
- The Iran-Iraq War began on 22 September 1980 and brought about a crisis that lasted until mid-January 1981.

All these crises had their origins in the Middle East, but they were rather different from each other in nature and in their consequences for Turkey.

The 1973 Oil Embargo

The crisis began in mid-October 1973, ten days after the outbreak of the Yom Kippur War. On 16 October the Organization of Petroleum Exporting Countries (OPEC) announced a unilateral increase in its benchmark price for oil (Arabian light) from U.S.$3.09 to $5.12 per barrel and also declared that from that time on, prices would fluctuate according to world market trends. The next day the Organization of Arab Petroleum Exporting Countries (OAPEC)[2] announced a schedule of production cuts based on an immediate 5 percent cut per producer, with further 5 percent cutdowns each month. Exports were to continue on a regular basis to all nations remaining friendly to the Arabs in their war and to those taking sanctions against Israel, but exports to the United States—Israel's chief arms provider—would be drastically reduced, with a view to stop completely the flow of Arab oil to that nation.

In effect, production cuts exceeded the objectives laid down in this schedule. On 19 October Saudi Arabia and Qatar announced a 10 percent cutdown in production (rather than 5 percent as called for by OAPEC), and OAPEC decided to impose an embargo on the Netherlands as well because of its traditional support of Israel. OAPEC members declared that a full embargo would also be imposed on any nation supplying oil to either the United States or the Netherlands. By the end of October 1973, oil production by the Arab members of OPEC had achieved 20 percent of the amount they were producing prior to the embargo (around 4 million barrels per day [mbd; one barrel of oil per day is abbreviated as bbd]), with Saudi Arabia taking the lead in production cuts.

On 22 December 1973 OPEC raised oil prices to a peak level of $11.65 per barrel. Just a few days later, however, the crisis began to fade when the organization announced its decision to reverse its previous policies and increase oil production by 10 percent across the board. Having reached the embargo's main economic objective—achieve the largest price hike possible—OPEC now began to relax its policies on production cuts. By April 1974 overall production was back to its previous level, and the crisis was over.

The economic ramifications of the oil embargo and the consequent steep price rises were drastic worldwide. The oil crisis contributed greatly to increased unemployment and inflation in both the Western world and developing countries. The posture taken by Arab oil producers during the crisis, dividing the world into friendly and hostile nations and diverting their oil shipments accordingly, did much damage to both Israel and its Western allies and forced many nations to declare support for the Arab cause in the Middle East conflict. Most African countries, for instance, severed diplomatic relations with Israel during that time. Even economically stronger nations, such as Japan, adjusted their foreign policies accordingly.

Turkey's economy in early October 1973 was fairly solid. Foreign currency reserves were approaching a record level of $2 billion, enough to cover Turkey's imports for nearly one year or its oil imports for three years (even at the new, quadrupled price). Of a total consumption of 10 million tons of oil a year, Turkey was producing 3.5 million tons domestically. The level of stockpiles in early 1974 was also reasonable, at about fifty days' consumption.

This economic stability, as well as the political aspects of the embargo—which on the face of it ensured that nations friendly to the Arabs would not be harmed—were enough to relieve Turkey of any worries, at least for the short term. The Turks did ask several producers to increase deliveries, but on the whole the new circumstances were not believed to warrant any drastic economic or political measures. Foreign currency reserves were drawn upon to finance oil imports at the new prices; Saudi Arabia and Iraq abided by their promise to keep Turkey supplied with oil during the crisis, and delays in deliveries from Kuwait were soon explained away as a "misunderstanding" and cleared out. Thus in 1974 Turkey encountered fewer technical difficulties in oil supplies than most Western oil importers.

Turkey was thus able to watch the rest of the world squirm, concerned only to have its leaders issue declarations of support for the Arab cause in the Middle East conflict, emphasizing that this was its traditional posture long before the crisis began. Typical of that policy was a statement by Foreign Minister Haluk Bayülken on 24 January 1974[3] in which he admitted that Turkey had encountered no difficulties throughout the crisis and expected the continued sympathy of the Middle Eastern oil producers because of its supportive attitude toward and close friendship with them. As we saw, Turkey began to feel the impact of the 1973 oil crisis only three years after it had ended, in spring of 1977, when it ran into serious problems financing oil imports and started experiencing fuel shortages.

The 1978 and 1980 Crises

In sharp contrast to its relative immunity during the 1973 crisis, Turkey was immediately and badly hit by the crises in both 1978 and 1980, which took place against a background of ongoing fuel shortages rooted in the 1973 crisis. The crises made Turkey's difficulties obtaining oil even more severe. The country searched frantically for alternative sources of oil while at the same time trying to find the credit needed to finance them. Bearing in mind the centrality of energy supplies to any national economy, this situation further exacerbated existing economic problems, which in turn led to social and political unrest (see Chapter 1).

Both crises began in a similar manner, with a cut in total OPEC output. In each case the cut amounted to around 2.5 mbd—about 8 percent of total world output. Turkey's oil supply was reduced by 44 and 55 percent, respectively, of its total amount of contract oil. All the same, production cuts were the only common denominator in the two crises.

The 1978 crisis was caused by domestic turmoil in Iran, beginning in late October. Until then, Iran had produced around 6 mbd, but the drastic reduction went on for several months (see Table 5.1), bringing about an almost complete cessation of Iranian oil exports until March–April 1979. The world energy sector was caught unprepared in terms of stockpiles, so the elimination of Iranian exports brought about a 100 to 150 percent increase in the price of oil during 1979. Once again,

Table 5.1   Iran and Non-Communist Oil Production Levels Before and After the Iranian Revolution (in mbd)

	Iranian Production	Non-Communist World Production
1978		
June	5,763	46,220
July	5,804	45,964
August	5,808	46,696
September	6,053	48,632
October	5,492	49,475
November	3,494	49,111
December	2,371	48,106
1979		
January	445	45,900
February	700	46,533
March	2,350	47,938
April	4,000	48,773
May	3,900	48,739

Source: Oil and Gas Journal, various issues, 1978, 1979.

unemployment increased in the developed countries, Western economic growth slowed significantly, and some Third World nations came close to bankruptcy.

The economic impact the crisis had on Turkey was enormous. Iranian oil constituted about half of Turkey's total crude oil imports at the time. Its elimination, grievous in itself, also disrupted Turkey's agreements with other oil producers. In some existing agreements the quoted price was increased; other producers simply defaulted on their contracts, as explained earlier. Thus Turkey was forced to purchase about a third of its oil imports for 1979 on the spot market at prices that sometimes reached $40 per barrel. These price hikes cost Turkey nearly $1 billion—an unbearably heavy burden for the Turkish economy at the time.

Additionally, fuel shortages throughout Turkey disrupted the normal course of life. Individual consumers were hoarding fuel, so even the expensive fuel brought in by the Bülent Ecevit government did not always reach the places where it could best serve the economy as a whole. Economic growth slowed, unemployment rose, and the education, health, and transportation systems were disrupted. With these economic difficulties came social unrest and domestic terror, which hastened the downfall of the Ecevit government. Toward the end of the crisis, in October 1979, Ecevit was replaced by a minority government headed by Süleyman Demirel.

By March 1980 the Demirel government had managed to overcome the devastating shortage of fuel, increase stockpiles, and accelerate the growth of industrial and agricultural production. Economic reforms introduced on 24 January 1980, in consultation with the International Monetary Fund and the World Bank, opened the door to Western economic assistance to Turkey. Also, as the rise in oil prices was halted, producers went back to fulfilling their contractual obligations, and Turkey had no further recourse to the spot market until the next crisis in the fall of 1980. Yet the temporary relaxation of economic stress did little to stabilize the tumultuous political system or stop domestic terror. On 12 September a military coup d'état took place, led by Chief of Staff Gen. Kenan Evren. Ten days later the military junta appointed a new government headed by retired admiral Bülent Ulusu. The main justification for this move was the political and social chaos that prevailed in the country.

Just one day after the new government came into power the Iran-Iraq War broke out, bringing with it the third global energy crisis. The flow of oil out of the two belligerent nations ceased nearly completely within a few days. As Table 5.2 indicates, the combined output of Iraq and Iran dropped from 5.6 mbd in August 1980 to 0.6 mbd in October. Other oil producers, especially Saudi Arabia, again increased their output as much as they could to fill the gap, so the quantity effectively withdrawn from the world market was about 2.5 mbd. Thus this crisis was similar to the previous one in terms of its impact on consumers.

Oil supplies to Turkey were again significantly reduced. At the outbreak of the war, Iraq was providing 40.2 percent of Turkey's total oil imports, and Iran contributed 17.0 percent more. Immediately upon entering office the new Turkish government had to find alternative sources for more than half of Turkey's oil imports. Furthermore, on 27 September 1980 Kurdish guerrillas hit the Kirkuk-Dörtyol pipeline, putting it out of commission. A few days later the USSR informed Turkey that because of the malfunctioning pipeline it could no longer deliver oil to Turkey.

Such a drastic cut in Turkish oil imports just as winter was approaching (Anatolian winters are notoriously harsh) put the newly appointed government and the generals behind it in a uniquely complicated position. Within two or three months Turkey's rulers had to obtain

Table 5.2   Iran and Iraq Oil Output, August 1980–May 1981 (in mbd)

	Iran	Iraq
1980		
August	1,300	4,300
September	1,100	2,900
October	450	150
November	700	300
December	1,200	550
1981		
January	1,200	400
February	1,500	700
March	1,800	960
April	1,600	800
May	1,600	900

Source: Petroleum Intelligence Weekly, various issues, 1980 and 1981.

about 3 million tons of oil to prevent yet another winter of discontent, similar to the one that caused so many problems in 1979–1980.

Turkey was the most seriously affected nation among Western oil consumers at the outbreak of the Iran-Iraq War. Other affected nations—such as Brazil, India, and Portugal—were depending on Iran and Iraq for 40 to 44 percent of their oil imports, whereas Turkey, as mentioned earlier, was importing 57.2 percent of its oil from those belligerents. Other importers, such as Spain and France, were dependent on Iran and Iraq for about a quarter of their total oil. Furthermore, Turkey had made barter agreements with the two nations, paying for its oil either in-kind or on credit; any alternative source would have required hard currency, cash on the nail. Turkey's balance of payments could not handle such terms of payment.

This was a particularly onerous test for Ulusu's new cabinet. The junta-appointed leaders were expected to restore law and order after many years of shortages and unrest, but they were facing a crisis that could have brought down even a popular elected regime. High hopes had been pinned on the results of the coup, but now it seemed doomed to grievous failure. In view of the new rulers' inexperience, as well as uncertainties about how long the crisis would go on, panic prevailed. Turkey was scrambling for oil, as well as for credit, from any possible source. It was a time of great anxiety.

The Generals' Struggle for Stability

The military coup on 12 September 1980, and the global energy crisis that broke out ten days later as a result of the Iran-Iraq War, fundamentally changed the political precepts whereby Turkey had been conducting both its domestic and foreign policy. The military regime had put the restoration of law and order at the top of its agenda. As noted, the need to combat political and social chaos was the chief cause of the coup and the main raison d'être for the new government. Thus these leaders could not afford a return to chronic shortages of essentials such as food, fuel, and medications or the collapse of education and transportation systems that had characterized the state of affairs the previous winter. Yet they immediately found themselves facing problems not only in financing oil imports but in getting them in the first place. Continued deterioration would have put the existence of the new government at risk, and no one knew where its collapse might lead.

Turkey's new government began frenetic efforts to find alternative sources of oil. The scope of the diplomatic campaign launched in late 1980 was extraordinary. Turkey made every effort to obtain whatever oil Iraq and Iran could produce and to induce the Libyans to double the amount already agreed upon. But above all it was looking for new sources, approaching Saudi Arabia, Kuwait, Abu Dhabi, Algeria, and Qatar, as well as the International Energy Agency (IEA) and the free market. A U.S. diplomat involved in the effort to help Turkey obtain oil in late 1980 spoke of the "panic" that reigned supreme at the Turkish Foreign Office during those months. The Turkish media spoke of a possible shortfall of 12 million tons of oil in 1981 and forecast a "black winter," just like the previous one. This scared the public and put even more pressure on the decisionmakers. Turkey's new leadership, determined to obtain all the oil it needed and to keep the social and economic systems intact, was ready to pay any price.

Saudi Arabia, it seemed to the Turks, was the only nation capable of helping Turkey obtain most of the oil it needed, and Turkey redoubled its efforts to reach some kind of agreement with that country. Turkey sought the help of Iraq, which was interested in Turkish neutrality in the war. Thus Iraq asked the Saudis to give Turkey the amount of oil it had lost because of the interruption of the Iraqi export. The Saudis were siding with Iraq at the time out of Arab solidarity vis-à-vis Iran, but they were still unwilling to consider a long-term agreement—something that was unattractive in terms of Saudi economic interests. Eventually, the

Saudis expressed readiness in principle to sell Turkey a small amount of oil as "emergency assistance" rather than under a bilateral agreement.

In London in early November 1980 Turkey signed a contract with the Texaco Corporation to import 320,000 tons of oil from Saudi Arabia and 110,000 tons from Kuwait by the end of 1980. This was the entire scope of the emergency assistance, and most of the oil had not reached Turkey when the crisis ended in mid-January 1981.

In October 1980 Turkey approached the IEA for the first time, asking for emergency assistance. Its first request was not specific enough, and in mid-December another request was made—to both the agency's Paris headquarters and to the State Department in Washington, D.C.—that was both specific and desperate.

IEA executives, having discussed oil consumption strategies for the duration of the war, decided in principle on 21 October to help Turkey obtain oil if shortages developed there as a result of the war. Preparations were made to secure 1 million tons of oil for Turkey by the end of 1980. The United States tried to help Turkey negotiate with the multinational oil corporations. But in view of Turkey's dire financial situation, those companies were totally uninterested.

Early January 1981 was a critical time for Turkey in its scramble for oil. A senior official of the National Petroleum Company was sent to Paris to present Turkey's dilemma to IEA leadership. The only result was a tentative expression of willingness by the Conoco Corporation to deliver a small shipment of Libyan oil—at 52 cents per barrel more than OPEC's posted price. But even that delivery could not be made before late January. The IEA effort to help Turkey was thus a failure. The agency was slow to react and then offered too little oil for too high a price.[4]

Saudi Pressure and Relations with Israel

Having withstood political pressures during the first two energy crises, despite the harsh consequences Ankara gave in this time. To an extent, the anxiety felt by every member of the new regime was responsible for the political caving in. On 2 December 1980, when the crisis was at its peak, Ankara announced its decision to lower the level of representation in its diplomatic relations with Israel and asked the chargé d'affaires and the military and economic attachés to leave the country. The new chargé, the Turks insisted, would be a lower-ranking official.[5] As a formal excuse the Turks cited Israel's 30 July enactment of a Basic Law defining Jerusalem as Israel's capital city, "eternal and undivided";[6] Jerusalem had been and remains the most controversial issue in the conflict between Israel and the Muslim world.

The political pressures Muslim oil suppliers brought to bear on Turkey in all three crises had to do mainly with its relations with Israel, an opportunistic intensification of the regular pressures expressed in various ways over the years. Throughout the 1970s the Turks firmly resisted all attempts to make the country sever formal relations with Israel, but in November 1980 the intended result was achieved, at least in part. In retrospect, it is ironic that the shortest global energy crisis in terms of duration, whose economic impact on the world at large as well as on Turkey was relatively mild, made Ankara feel cornered. It therefore took a step it would likely not have taken of its own free will.

The downgrading of Turkey's relations with Israel was the most drastic political move in the process of political rapprochement between Turkey and the Muslim nations in the period 1973–1981. Until then this process had advanced gradually, as part of an overall trend in Turkish foreign policy dating back to the mid-1960s. Even the intensive economic relations developed with Muslim oil producers in 1978 and 1979 did not cause Ankara to lose control over its political orientation. Turkey was careful to maintain good relations with the United States and the rest of the Western world, taking policy cues from other Muslim nations only in matters that could not possibly antagonize the Americans. But the decision to lower the level of relations with Israel was taken swiftly and secretly, a step Turkey knew in advance would bring about strong protests by the United States and the rest of the Western alliance.

In 1974, when Saudi Arabia leveled political and Islamic pressure through Lütfi Doğan and Necmettin Erbakan, public reaction was fierce, indicating Turkey's ability at the time to withstand such pressure from the oil producers. In 1980, however, circumstances were completely different. The seemingly endless economic crisis had weakened the nation, as its various systems crumbled for lack of fuel. Then came the fear that fuel would be impossible to obtain at all, and the highest political leaders found themselves constrained to take drastic action. Furthermore, Ankara feared Western assistance might be reduced because of the antidemocratic coup.

In reviewing the impact of the earliest stages of the Iran-Iraq War on Turkey, a sharp contrast stands out between the feared loss of nearly

60 percent of the contract oil promised and the actual political and economic effects the crisis had on Turkey. Both Iran and Iraq needed
Turkey to remain neutral in the war between them, as well as to provide
them with export goods essential for their economies. Therefore both
nations courted Turkey intensely after the war started. They were relying on Turkey as a land route for the inflow of commodities and equipment from the West, because the war effectively cut off their sea routes.
Besides more Turkish exports, they also needed Turkey's cooperation in
curbing the activities of Kurdish separatists. Consequently, both nations
were anxious to keep up their oil deliveries to Turkey, even as the war
raged on. The trouble was that in the early months they were physically
unable to do so (see Table 5.2). They were also unable to exert any
political pressure on Turkey, especially on matters not directly related
to the war.

Libya was also in no position to bring political pressure to bear on
Turkey during the 1980 crisis. Once again, Libya's commercial interests
were in conflict with its desire to move politically closer to Turkey.
During October–December 1980 it was unclear how the global crisis
might proceed, and Libya could not afford to increase its contractual
obligations for fear of losing money if prices went up again. Therefore
Libya turned down Turkish pleas, albeit reluctantly, and thus was in no
position to make any politically controversial demands. Its hostility
toward Israel, it seems, did not outweigh its fear of losing possible
windfall gains.

With Saudi Arabia, however, things were very different.
Throughout 1980 Turkey had endeavored at the highest political level
to obtain a loan from that country to help ease its financial difficulties.
It had made a formal request for a $1 billion loan when Finance
Minister İsmet Sezgin visited Riyadh in December 1979. This request
was repeated in both March and May 1980, first when Sezgin returned
and then during a visit by Turkish head of planning Turgut Özal.
Eventually, the Saudis announced their decision to arrange a $250 mil-
lion loan for Turkey. The Saudi cabinet approved the loan, and the
appropriate royal decree was issued. In July 1980 a Saudi delegation
headed by Minister of Finance Shaikh Abu Khail was to arrive in
Ankara to sign the loan agreement. In July and August 1980, however,
the shaikh's visit was postponed at least four different times. He finally
arrived on 23 August, one month before the Iran-Iraq War broke out,
and the agreement was signed.

But when the war began on 22 September 1980, $150 million of the

$250 million had not yet been transferred, and the Saudis withheld further transfers until the situation cleared. Not until 26 November 1980, about two weeks after Foreign Minister İlter Türkmen visited Riyadh, did Radio Ankara announce that the unpaid portion of the promised credit would be transferred on 2 December. In other words, Saudi credit arrived at Turkey's Central Bank precisely on the date the government announced its decision to downgrade relations with Israel.

Saudi Arabia—unlike Iraq, Iran, and Libya—was able to exert political pressure on Ankara at the time for several reasons. Its production capacity remained intact and had even expanded since the outbreak of the war to take up some of the slack created by Iran and Iraq. It also had huge financial reserves, so making a loan to Turkey was no problem. And unlike Iran and Iraq, Saudi Arabia was not dependent on Turkish goodwill economically, politically, or strategically.

The oil panic created by the Iran-Iraq War made it even easier for Saudi Arabia to pressure Turkey. Saudi promises to increase oil supplies (none of which could have been realized before the end of 1980) and the suspension of an already contracted loan put an edge on the triple demands the Saudis made to the Turks: to sever or at least downgrade relations with Israel; to become more involved in Muslim affairs, that is, less Western oriented culturally; and to allow Saudi activities within Islamic projects both inside Turkey and overseas.

The Turkish government believed meeting Saudi demands in terms of both relations with Israel and Islamic affairs would go a long way toward relieving Turkey's dangerous situation. Relations between Turkey and Israel were already tense in light of Turkey's reaction to the Jerusalem Law, and further deterioration seemed likely. Thus Turkey described its decision to downgrade the level of representation as a further and stronger response to that legislation, unconvincingly explaining that Turkey had delayed its reaction because it had hoped Israel would reconsider its decision following a severe reprimand by the UN General Assembly.[7]

Obviously, this rationalization was far removed from Turkey's actual motives. Hayrettin Erkman, foreign minister at the time the Jerusalem Law had been adopted, described the situation thus: "By recalling its ambassador from Tel Aviv and closing down its Consulate-General in Jerusalem in late August 1980, the Demirel government had already reacted to this legislation. This constituted the entirety of our response, and I do not think the prime minister or myself intended any further action. After 22 September 1980 [the outbreak of the Iran-Iraq

War] there already were other factors and other pressures that produced the decision to lower representation."[8] These "other pressures" referred to the new circumstances created by the war.

All of this notwithstanding, it seems no secret agreement between Saudi Arabia and Turkey regarding the immediate provision of oil or money lay behind Turkey's November move against Israel. Rather, the new policy was based on a general assumption prevalent among decisionmakers in Ankara, particularly in the Foreign Office, that the Saudis and other Muslim oil producers would react favorably to such a move. In other words, the Turks allowed their policies to be guided by wishful thinking, perhaps the best indication of the state of panic that prevailed in Ankara in the winter of 1980–1981.

In fact, the decision to downgrade the level of representation failed to bring about any change of attitude in Riyadh, at least regarding oil. Only in March 1981—four months after the move against Israel and two months after the energy crisis had ended—did the Saudis express willingness to sell oil to Turkey. By that time, however, the Turks were no longer desperately scrambling for oil. The requests made in November and December 1980, to purchase 5 million tons of oil from Saudi Arabia, were withdrawn. Turkey was able to overcome the oil shortages in January 1981 because the global crisis was over, not because of the action it took against Israel.

Ironically, Ankara, having withstood nearly ten years of repeated pressure to sever its relations with Israel, had given in to such political pressure just a few weeks before it finally managed to lay to rest the oil import problems that had plagued it since the summer of 1977. By the end of February 1981, when Israeli chargé d'affaires Ya'acov Cohen left Ankara, Turkey had oil surpluses and was even able to resume the importation of oil distillates.[9] This amazing turnaround, from an oil panic up to mid-January to a state of surplus just a few weeks later, casts serious doubts on the advisability of damaging Turkey's relations with Israel and submitting to Saudi demands on Islamic affairs.

Notes

1. See Eli Arom, *IEA and the 1979 Oil Crisis Management* (in Hebrew) (Tel Aviv: Israel Petroleum Institute, 1980), p. 4.

2. At the time, OAPEC included Saudi Arabia, Kuwait, Libya, Algeria, Dubai, Abu Dhabi, Bahrain, and Iraq.

3. Anatolia News Agency, 24 January 1974.

4. By February 1981, when the IEA aid was scheduled to arrive, Turkey had solved its problems by itself through its traditional suppliers.

5. Consequently, I, then second secretary in the Israeli Foreign Office, was sent to represent Israel in Ankara.

6. Israel has no written constitution; instead, from time to time it enacts Basic Laws on matters of great importance.

7. In August 1980 the UN gave Israel a deadline, 15 November, to revoke the Jerusalem Law.

8. Interview with the author, Ankara, 7 May 1985. In fact, Erkman announced publicly on 28 August 1980 that the closure of the consulate in Jerusalem would have no effect on the level of representation in Turkey's relations with Israel. Prime Minister Demirel, when I interviewed him in Ankara, also indicated that the August 1980 reaction had constituted Turkey's entire reaction to the Jerusalem Law.

9. The first announcement of Turkish oil products exports was broadcast by Radio Ankara on 4 February 1981.

6

THE 1980S:
HAPPY DAYS ARE HERE AGAIN

During the early 1980s the economic base of relations between Turkey and the Muslim countries was greatly expanded to include more than just trade. The main trends during the 1970s—increased purchasing power and ambitious development plans for the oil producers on the one hand and a decline in the economic growth of Western nations on the other—diverted Turkish human resources and entrepreneurship toward the producing nations. The economic slump in the West brought the flow of Turkish laborers to Europe to a near standstill and affected—not favorably—the flow of Western foreign aid to Turkey and the hard currency transfers from Turkish laborers still employed abroad. The combined effects of all these factors in terms of Turkey's balance of payment made it imperative for Turkey to find new sources of foreign currency revenues. The oil producers were the only realistic possibility, because only they seemed able to absorb Turkish export goods, laborers, and know-how in large quantities.

According to Turkey's Employment Service figures, 93,843 Turks went to work overseas between 1978 and 1981: 33,719 (36 percent) went to Libya, 30,993 (33 percent) to Saudi Arabia, and 13,950 (about 15 percent) to Iraq.[1] Thus nearly 85 percent of the Turkish labor force that went to work abroad during those years went to oil-producing nations, whereas Germany—still the largest Western employer of Turkish Gastarbeiter—absorbed only 3.5 percent of the total in 1978–1981. By 1982 the number of Turks working in Arab countries had reached over 130,000—80,600 in Libya, 35,000 in Saudi Arabia, and 15,000 in Iraq. In 1982 alone those laborers were paid $810 million in hard currency. During the rest of the decade, the number of Turkish workers in Arab countries grew still larger, reaching about 300,000 by 1990; of those, 155,000 were in Saudi Arabia, 50,000 in Libya, and 12,000 in Iraq.[2]

The increase in the number of Turkish contracting companies operating in the oil countries was even more impressive. In 1978 nearly 20 Turkish companies were functioning in several Arab countries, under contracts amounting to a total of $1,650 million. About half were doing construction work in Libya—building or repairing harbors, roads, factories, and apartment buildings—and the rest operated in Saudi Arabia (factories and mosques), Kuwait, and Iraq. By 1981 the total number of such companies had reached 113 (including 68 in Libya, 19 in Saudi Arabia, and 13 in Iraq). The number grew to 232 in 1982 (including 98 in Libya, 79 in Saudi Arabia, and 35 in Iraq) and to 283 in 1983 (including 109 in Saudi Arabia and 105 in Libya). The total dollar value of these companies' contracts had reached nearly $13 billion in 1982 ($8.2 billion in Libya, $3.3 billion in Saudi Arabia, and nearly $1 billion in Iraq). This sphere of activity expanded further over the next few years, although at a slower rate. The figures for 1988 indicate an overall contract value of almost $17 billion: $9.4 billion in Libya, $5 billion in Saudi Arabia, and $2 billion in Iraq.[3]

As the numbers of both Turkish laborers and Turkish companies operating in the oil countries kept growing, Turkey became increasingly dependent on foreign currency transfers from those sources. In 1981 and 1982 unilateral transfers by workers alone financed nearly a third of Turkey's oil bill.

Turning the Tables

The 1980s were years of economic stability for Turkey, marked by rapid economic growth and relief from oil import constraints.[4] Beginning in 1981, global oil prices gradually decreased. The Organization of Petroleum Exporting Countries' (OPEC) posted price was $34.50 per barrel in 1981, but the price kept going down until it reached $17.60 per barrel in 1989.[5] For Turkey, which was spending 40 to 50 percent of its total import bills on oil in the early 1980s, this decrease brought significant relief.

When the oil crisis was at its height in 1979–1980, Turkey purchased most of its oil on the spot market, paying much more than the posted price; thus the decline in oil prices was even more welcome than the figures apparently indicate. In 1980 Turkey paid $3.9 billion for 10.5 million tons of oil, whereas in the late 1980s it was paying a much smaller sum for double the amount of oil (needed because of its rapid economic expansion). For the 21.7 million tons of oil imported in 1988 Turkey paid $2.4 billion,[6] and the figures for 1989 were similar in terms of both quantity and cost; that year Turkey paid 25 percent less in dollar terms for nearly double the amount imported in 1981.[7]

The decline in oil prices had far-reaching consequences for Turkey's economic relationships with the Muslim oil producers, which remained nearly exclusive suppliers of Turkey's oil. The oil revenues of nations such as Iraq (still Turkey's chief supplier), Iran, and Libya declined, causing financial strains in all three economies. This badly hurt their ability to pay both for Turkish exports and for the work done or intended to be done by Turkish contractors in their territories.

The situation had reversed itself. Turkey was relieved of its previous finance troubles, whereas Iraq, Iran, and Libya were experiencing such difficulties for the first time in their histories. Ankara-Baghdad relations were now clouded by an accumulating Iraqi debt, which reached nearly $3 billion toward the end of the Iran-Iraq War. In April 1988 Turkey threatened to cancel Iraq's commercial credit, receiving some repayment as a result, but in mid-1989 Iraq still owed Turkey between $2 and 2.5 billion.[8] Iraq was in dire financial straits indeed, having accumulated during the war an overall external debt amounting to around $80 billion,[9] and Turkey relented, agreeing to open a credit line for Iraq extending to $1.3 billion in 1989.[10] Of this, $400 millions was earmarked for Turkish export goods. At the time, Western banks, fearing imminent Iraqi bankruptcy, were refusing to make loans to Baghdad unless they were guaranteed by foreign governments (such as the United States or Japan).

Despite Iraq's financial difficulties and the bitter dispute between the two governments over debt repayment, commercial relations went on much as usual during most of the decade. Iraq kept pumping oil to Europe through Turkey, reaching a volume of 1.6 million barrels a day in the late 1980s. Bilateral trade amounted to around $2 billion a year during the 1980s, even though its share of total Turkish trade gradually diminished toward the end of the decade.

Turkey's economic relations with Libya in the 1980s were much less stable, deteriorating rapidly during the second half of the decade. In the mid-1980s severe disputes erupted between the two countries over the price of Libyan oil, as well as arrears in payments to Turkish contractors. In 1989 Libya's accumulated debt to Turkish contracting companies was estimated at $2 billion, including $400 million to twenty companies for work already completed.[11] These defaults drove many Turkish construction companies to declare bankruptcy, and Turkish laborers staged hunger strikes over unpaid years of work in Libya.

As a result of this dispute, an estimated $1.5 billion in Libyan projects to be done by Turkish companies were suspended, and the volume of bilateral trade declined. In 1988 Turkish-Libyan trade amounted to only $297 million, compared with $1.2 billion in 1981. Oil imports from Libya were almost completely terminated; in 1988 Turkey imported only 300,000 tons of oil from that country, less than the amount it imported from China. In an effort to mend fences, in 1989, after very draining negotiations, Turkey agreed to open a $300 million credit line for Turkish imported goods if Libya would pay for the rest of its imports in cash (rather than oil, which Turkey no longer needed). Also, the Turks demanded that Libya increase the advances made to Turkish companies in Libya while work was still in progress.[12]

Similar disputes also arose between Turkey and Iran during the second half of the 1980s. And again, disagreements over the price of oil led to a reduction in the volume of trade during that period. Again Turkey offered credit, this time amounting to $700 million—$400 million for joint projects, the remainder to finance Iranian importation of Turkish goods.[13]

Thus the end of the global energy crisis and the decline in oil prices brought about a complete turnaround in economic relations between Turkey and the Muslim oil countries. After four desperate years (1977–1981), a time when Turkey badly needed oil import financing from Iraq, Iran, and Libya, the situation was reversed: during most of the late 1980s those oil producers owed Turkey a total of around $5 billion. Turkey still needed economic assistance, but that now came from the West, especially the United States. In fact, Turkey now ranked third among U.S. aid recipients, after Israel and Egypt (nonmilitary aid to Turkey stood at $720 million in FY 1986–87 and $525 million in FY 1987–88).

Trade with the Muslim World During the War

Trade figures indicate a decline in the financial volume of Turkish bilateral trade with the Muslim world in the late 1980s. In 1981 the overall volume of trade between Turkey and Middle Eastern Muslim nations reached $5.4 billion, increasing to $6.2 billion in 1983 and again to a peak of $6.9 billion in 1985; by 1987 it had fallen to $5 billion. In relative terms, the decline was even more pronounced. Turkey's trade with these nations accounted for 40 percent of its total foreign trade in 1981 and nearly 42 percent in 1983, but in 1985 it amounted to 36 percent of the total, dropping to under 22 percent in 1987. From then to the end of the decade, trade with the Muslim countries consisted of one-fifth to one-quarter of the total.[14]

This picture, however, does not truly reflect the benefits Turkey derived from its commerce with the Muslim world. A major part of the decline was accounted for by the collapse of oil prices, which made Turkey spend less on its imports from those countries while increasing its exports to them. The balance of trade between Turkey and the Muslim nations thus tilted during the 1980s from a state of significant deficit in the 1970s (with exports covering just 10 to 20 percent of imports) to parity in the mid-1980s, moving well into the black toward the end of the decade. In 1988 Turkish exports to Muslim nations outweighed imports from them by $1.5 billion.

The Iran-Iraq War opened up new commercial opportunities for Turkey and played an important role in the rapid expansion of its export sector. From $5 billion in 1981, total Turkish exports rose to $12 billion in 1988. As the war raged between them, Iran and Iraq became highly important destinations for Turkish exports. Annual trade with Iraq during most of the duration of the war remained at $2 billion, increasing to $2.5 billion in 1989. Exports to Iran reached about $2.3 billion in 1983, dropping to less than half that in 1988. But even then Iran remained Turkey's second-most-important Muslim trade partner. In all, for most of the decade Turkey's trade with Iran and Iraq constituted 50 to 60 percent of its overall trade with the Muslim world.

At the same time, Turkey was improving its trade balance with other Muslim nations as well. Between 1985 and 1988 Turkish exports to Saudi Arabia amounted to two to four times its imports from that country.[15] Trade with Kuwait (about $350 million in 1988) and Algeria (about $200 million that same year) became significant during the

1980s. Trade with Egypt, Jordan, and Lebanon maintained a total volume of $150–$200 million a year, whereas annual trade with Syria did not exceed $75 million for most of the decade. The balance of trade with Kuwait, Algeria, Egypt, Jordan, Lebanon, and Syria was tilted very heavily in favor of Turkey.

Water Power

As related earlier, water provides Turkey with one of its greatest economic and political advantages in its relations with its Middle Eastern neighbors. The waters of the Euphrates and Tigris Rivers, flowing out of Turkey to Syria and Iraq, have become a fiercely contested issue with both Damascus and Baghdad as a result of extensive development plans drawn for southeast Turkey. These plans—many of them already completed—have included the construction of many dams and power stations along the two rivers. The overall Turkish objective is to make this part of the country the grain basket of the Middle East, as well as to satisfy more than half of Turkey's energy requirements with hydroelectric power.

During the 1980s Turkey gradually increased its control over the rates of flow of both the Euphrates and the Tigris, and now it has nearly complete control. In 1990, when the Turks set about filling the Atatürk reservoir—the largest on the Euphrates, about 65 kilometers (km) north of the Syrian border—they reduced the flow of water downriver for quite some time. Protests by Iraq and Syria about the damage to agriculture and industry were to no avail, testifying to Turkey's stranglehold over both countries.

At the same time, as noted earlier, Turkey had proposed ambitious projects that could provide significant amounts of water to its arid and semiarid neighbors to the south. In the mid-1980s Prime Minister Turgut Özal announced his ambitious "Peace Pipeline" project, consisting of two pipelines intended to stretch from the Sayhan and Ceyhan Rivers, near Adana, all the way to the Arabian Peninsula. The western pipeline would provide 3.5 million cubic meters of water to Syria, Jordan, and western Saudi Arabia; and the eastern branch would deliver 2.5 million cubic meters to Kuwait, eastern Saudi Arabia, Bahrain, Qatar, the United Arab Republic, and Oman.

For various economic and political reasons, this grandiose project did not materialize, but Özal did not give up. In 1990 he came up with a new idea—to ship Turkish water abroad in tanker ships. International trade in water is uncommon, but Özal earmarked the water of the Manavgat River, which flows to the sea about 60 km from the town of Antalya, for this project. It took Turkey ten years to build a sophisticated purification and loading facility there. Its offshore terminals allow for the loading of supertankers.

At the time of writing, Turkey was ready to export 200 million cubic meters of water per year, but it still faced two problems: customers and suitable tankers. Tankers have never been used or intended to be used for transporting drinking-quality water, so the ones to be used will have to be refitted at a cost that will inevitably affect the price of water for customers. With the cost of desalinated water now ranging between eighty and ninety cents per cubic meter, Manavgat water must be offered at a lower price to compensate for the political cost to any customer of becoming dependent on a foreign nation (that is, Turkey) for this most vital liquid.

The idea is being examined seriously by Israel, Jordan, and the Palestinian Authority, the last two of which, however, would have to rely on Israel for unloading facilities. In view of the complicated and unstable relationships between these countries, the notion of creating a partnership for water importation from Turkey—in spite of their desperate need for more water—is fraught with problems, although such a venture could stabilize their relationships. Negotiations for shipping Turkish water to Israel itself are at a progressed stage.

The 1990s—Back to the West

As of 1990 Turkey's economic ties with the Muslim world had even less importance than previously. Still, Turkey was importing most of the oil it needed from Muslim nations; and Muslim markets, so close geographically, provided convenient outlets for Turkish industrial exports. As ever, Turkey's geographical position still made it an important route for land transport between the Middle East and Europe, as well as a channel for oil to Mediterranean ports. The turn to the east during the 1970s had a certain economic rationale, even though Turkey had little choice in the matter at the time. This trend persisted for most of the 1980s, as Turkey gradually shrugged off the economic dependence that had characterized its relations with the oil countries.

Nevertheless, under the leadership of Turgut Özal (1983–1992),

Turkey's economy gradually reasserted its Western orientation. Turkish exports, which had increased rapidly, were again directed westward.[16] In 1995 the combined share of all Muslim countries in Turkey's exports amounted to 13.4 percent—down from nearly half the total export ten years before. That same year, Organization for Economic Cooperation and Development countries accounted for 61.1 percent of total Turkish exports. Thus most imports, with the decline in the proportional share of oil, were coming from the West, especially from European Union (EU) countries. In all, the fast growth of Turkey's foreign trade (at a rate of about 15 percent per year in the 1980s and 1990s) clearly indicated the globalization and modernization of Turkey's economy during that period.

In January 1996 an EU customs union went into effect, and Turkey intensified its efforts to become a full EU member. That goal has guided its foreign policy and economic priorities since that time. Turkey's efforts were dealt a severe blow in July 1997 when the European Commission, in a report entitled "Agenda 2000," failed to include Turkey in the list of eleven nations considered candidates for full membership in the near future. Yet Turkey persisted in its efforts, which were rewarded two years later. At their meeting in Helsinki in December 1999, the EU foreign ministers finally decided to make Turkey a candidate for membership. It may take at least a decade for Turkey to become a full member of the EU, but the Turkish economy is gearing up for this new status, which would anchor Turkey once and for all to the Western economy.

These changes, hurling Turkey from East to West, have affected Turkey's international politics in the Middle East and beyond, the subject of Part 2 of this book.

Notes

1. As regularly published by the Turkish Economic News Agency.

2. *Turkish Daily News,* 14 May 1990.

3. These data are based on the Turkish State Institute of Statistics, *Statistical Yearbook of Turkey* (Ankara: Government of Turkey, 1989), and annual publications of the Turkish Energy Office.

4. In 1987 alone, the gross national product (GNP) grew by 7.4 percent (*Middle East Economic Digest [MEED],* 28 July 1989).

5. CIA, *International Energy Statistical Review* (Washington, D.C.: CIA, 27 January 1987 and 27 February 1990).

6. Turkish State Institute of Statistics, *Statistical Yearbook of Turkey.*

7. *Cumhuriyet,* January 1990; *Turkish Daily News,* 18 January 1990.

8. *MEED,* 17 March 1989; Radio Ankara, 31 May 1989. Some Turkish sources spoke of a debt as high as $2.7 billion.

9. *Turkish Daily News,* 18 January 1990.

10. *MEED,* 17 March 1989.

11. *Hürriyet,* 25 August 1987; Reuter's, 21 September 1989.

12. *Financial Times,* 10 January 1989; Reuter's, 21 September 1989.

13. Iran News Agency, 2 March 1990; also AP (from Nicosia), same date.

14. International Monetary Fund, *Directions of Trade* (Washington, D.C.: IMF, 1989).

15. Toward the end of the 1980s, following disputes with Libya and Iran over the price of oil, Turkey again increased the amount of oil it imported from Saudi Arabia.

16. Turkish State Institute of Statistics, *Annual Foreign Trade Statistics* (Ankara: Government of Turkey, 1985–1998).

PART TWO

TURKEY IN THE MIDDLE EAST:
INTERNATIONAL ASPECTS

7

THE MUSLIM OPTION

As Turkey was turning toward the Middle East in its economic policies during the 1970s and 1980s, the ideologies upon which it had traditionally based its foreign policy had to be reexamined. How should Turkey feel about the new economic strength some Muslim nations had gained? What were the relative weights of ideology on the one hand and basic economic needs on the other in this policy realignment?

Turkey's Attitude Toward OPEC

Just one day before assuming office as prime minister, Bülent Ecevit had made abundantly clear his sympathy with the oil producers' actions during the 1973 embargo. In a column in *Ankara Bayram* on 6 January 1974, Ecevit made his views known: "The only kind of nationalism still valid nowadays is economic nationalism, as proven by the events of these last few months. The oil shaikhdoms of the Middle East are now way ahead of Turkey in understanding the concept of nationalism and adjusting it to contemporary realities and requirements. The Turkish people, the first one [in the Middle East] to raise the banner of nationalism fifty years ago, now lags behind those nations that in the past had regarded it as a pioneer in understanding modern nationalism." Ecevit described moves by Organization of Petroleum Exporting Countries (OPEC) members to seize control over their oil assets as the appropriate economic model for those times, which should serve as a model for Turkey as well.[1]

Only "economic nationalism," he said, could prevent the exploitation of a developing country by outside forces: "We saw clearly what has happened to countries that have placed their oil production in foreign hands, running their energy and industrialization policies according to foreign advice." The remarkable support for the oil embargo shown by Necmettin Erbakan, chairman of Milli Selamet Partisi (National Salvation Party), was accompanied by much optimism about its implications for Turkey. One day before assuming office as deputy prime minister, he, too, published his views on the subject: Turkey would benefit from the energy crisis, because it could obtain as much oil as it required at "appropriate prices"; that would enable Turkish industry to operate at full capacity and even expand. Leveraged by industrial products manufactured using cheap oil, Turkey could occupy a place of importance in Western markets, as well as secure for its products significant market shares in the oil-producing countries. "This is a once-in-a-lifetime opportunity. A clever and thoughtful government could make a giant leap forward using the conditions now created."[2]

Having concluded that the energy crisis would increase Arab purchasing power and open new, important markets for Turkey, the Demokratik Parti (Democracy Party), which controlled one-tenth of the seats in parliament, decided to change its position on the issue. Two of its leaders, Vedat Ünsal and Faruk Sükan, called on the government to shift Turkey's economic orientation from the European Common Market toward the Muslim nations of the Middle East. Since Europe could no longer provide proper markets for Turkish goods, the protocol already signed with the European Economic Community (EEC) should be revoked and economic relations with the Muslim nations developed instead.

Finance Minister Deniz Baykal expounded on the new doctrine at a press conference on 30 August 1974:

> Our term of office in government began at a time when the developing nations first launched their struggle to undermine the existing economic balance and remove accumulated distortions. The past twenty-five years failed to fulfill the hopes of the developing nations to narrow the gap between themselves and the developed nations. The former have always found themselves selling their goods cheap and buying expensive finished products from the West. The industrialized nations have always been the big winners in international trade. Using

cheap raw materials and energy resources, the developed nations have been able to keep prices at a low level domestically, passing on the burden of inflation to the developing nations.

Four years later Prime Minister Ecevit was disillusioned but unrepentant. As he told a conference on the new international economic order (NIEO) in Istanbul, "The move made by the oil producers in late 1973 had raised hopes for breaking the vicious circle of the widening gap between developed and developing nations. These hopes were frustrated because much of the capital accumulated by the oil producers was channeled into the banking institutions of the existing order, and still more was used for real estate investments in the developed countries."[3]

Not a few leaders of the coalition that led Turkey until September 1974 were naively sympathetic toward the moves made by the oil producers. Still, most were more realistic than Erbakan in viewing the implications for Turkey. Ecevit told the German TV network UPI that the serious situation in which the developing countries found themselves following the energy crisis would only grow worse. He expressed his hope that the oil producers would not choose a path that would make it even more difficult for nations such as Turkey to develop their economies. Foreign Minister Turan Güneş told *Cumhuriyet* that in his view, the least developed and developing nations would emerge from the energy crisis as the big losers, whereas the industrialized nations—despite their dependence on OPEC oil—would find solutions to the problems raised by the crisis.[4] He added that Turkey was going to lose $800 million in 1974 alone because of the oil crisis.

As 1974 drew on, the conflict of attitudes became clearer: despite its instinctive sympathy toward the assertiveness now shown by the oil producers, Turkey was fast becoming one of the main victims. "There is a contradiction here between our practical interests—cheap oil—and our philosophy. The way I see it, practical interests must never override values," said Minister of Finance Baykal.[5]

Not surprisingly, the Adalet Partisi (AP, Justice Party), then in opposition (the party returned to government in April 1975), took a different view. The AP's leader at the time, Süleyman Demirel, said: "I don't care who gets richer as a result of the energy crisis. What I care about is that we are getting poorer. It is the oil producers who are becoming richer, not I. I do not run the 'producers' or the 'developing' nations, I run Turkey. I figure out how much I buy and how much I pay.

If a barrel of oil costs twenty-five dollars—to think that we stand to gain anything from it indicates a kind of Robin Hood mentality." Demirel added: "When oil prices went up, the industrialized nations raised the prices of their products. We gained nothing from the oil price hikes or from the Western industrial product price hikes, because we had to purchase these, too. Turkey was robbed twice. We needed more money just to survive. How can this be good?"[6]

Economic Integration in the Middle East

The leaders of the ruling Cumhuriyet Halk Partisi (CHP, Republican People's Party), while supporting OPEC as a matter of principle, also sought to strengthen Turkey's economic and political ties with the Middle Eastern oil producers. They expected the energy crisis to increase the power of the nonaligned bloc, in which Turkey could play a major role without relinquishing its political and military connections with the West.

In their efforts to align Turkey with the oil producers, they enjoyed the full support of their coalition partner, the Milli Selamet Partisi (MSP, National Salvation Party). Between 1974 and 1978 the MSP held several major government portfolios, including the Ministry of Industry. Led by Erbakan, the MSP regarded the Middle East as Turkey's "spiritual sphere of existence," as well as its natural economic setting. Erbakan undertook as a holy mission the integration of Turkey, culturally and economically, in the Middle East. He wholeheartedly believed the "Muslim option" could and should replace Turkey's reliance on the West and even called for an Islamic common market and military alliance.[7]

Ecevit developed the notion of economic integration in the Middle East soon after the 1973 oil embargo, and he adhered to it consistently throughout the 1970s. Addressing parliament on 3 July 1974 he said that economically, great opportunities for cooperation between the West and the Middle East had come to the fore and could be advantageously exploited, provided Middle Eastern countries joined forces: "Should any Middle East nation, including Turkey, decide to make separate economic ties with the West, this would be interpreted as a sign of weakness. But if Middle Eastern nations could unite to undertake projects that none of them could carry out alone, and bring their national economies together in harmony, they would achieve a firmly based economy, which should make it easier for them to cooperate with the West."[8]

Even as opposition leader, Ecevit stuck to his guns. While visiting Yugoslavia in 1976 he told an interviewer: "No single underdeveloped or developing nation is strong enough to safeguard its survival and unique character vis-à-vis the great powers. Nations whose economies are complementary must create groups of economic cooperation in each region. To achieve this, steps must be taken to promote economic integration in the Balkans, in the Middle East, and to an extent, in the entire Mediterranean area."[9]

Ecevit hoped to bring this notion to bear on Turkey's relations with Libya, for instance: "Such a solidarity is requisite in order to maintain our independence against abuse by the imperialistic movements that keep on going under different names and in different forms. The relations of solidarity between Turkey and Libya ought to set an example for other developing nations."[10]

Ecevit's notion of integrating Turkey into the Middle East was largely derived from his overarching view of the Third World.

> We do not propose to refrain from economic cooperation with the developed world. But we do believe that by joining forces, the developing nations in this region, including Turkey, could open up to the world more readily. Even though there are obstacles on the road to increased cooperation between developing nations, it must be borne in mind that the expanding gap between them and the developed nations must lead eventually to an economic and social crisis, such that the developed world too would not be able to escape its consequences.[11]

Ecevit's notion of bringing Turkey economically and socially closer to Islamic and Third World nations was complemented by a political vision that combined seeking a more significant role for Turkey in those parts of the world with a more pluralistic and equitable global political order. In his view, a rapprochement between Turkey and the Third World, including the Muslim oil producers, was a new "option," capable of counterbalancing Western influence in the region and yielding economic benefits to its inhabitants. Adopting this "southeastern option" did not necessarily mean abandoning the Western option, but it did involve wholehearted Turkish support for the new international economic order and an effort to obliterate Turkey's image as an ally of the West in the eyes of Third World nations: "If Turkey wishes to contribute to the accelerated development of the oil producing nations and bolster its political position in the Middle East, it must not be portrayed as a representative of the West amongst Middle Eastern nations."[12] Later, Ecevit went even further: "By bringing about full solidarity among the various nations in the Middle East, we shall make them all the more capable of withstanding machinations, collusions and provocations designed to weaken the region as a whole, and any single country in it."[13] The general idea was to reduce as much as possible the great powers' involvement in Middle Eastern affairs.

Observing relations with the West from a Middle Eastern perspective was a novel concept in Turkish political thinking, emerging as an early consequence of the energy crisis and the increased economic might of Middle Eastern oil producers. CHP leaders feared Turkey's relationship with the West could jeopardize its relations with the Arab countries. The precepts of the NIEO[14] and the change in Turkey's image in the Third World, including among the oil producers, should have provided the basis for rapprochement with those nations.

The 1974 Cyprus War and the vehement reaction of the West (especially the United States) to Turkey's moves there highlighted the uncertainties inherent in Turkey's global orientation. Yılmaz Altuğ, a professor of international law who later became a member of the Foreign Affairs Committee of the Turkish parliament, published an article in which he concluded that Turkey had only two options in the international arena: reliance on the West (excluding the United States because of its reaction to Turkey's involvement in Cyprus) as the preferred option or reliance on Muslim oil producers as a second choice.[15] That ranking reflected Altuğ's view that the oil producers were incapable of helping Turkey solve its defense problems. Even petrodollars, he said, could not ensure the flow of the spare parts needed to maintain Western-made weapon systems; hence reliance on European North Atlantic Treaty Organization (NATO) members was the best choice.

The Southeastern Option

The position developed by the CHP on the energy crisis and the NIEO was clearly reflected in Turkish foreign policy during the period 1974–1981. Ecevit laid down the law on these issues, and Turkey's policies vis-à-vis the Third World, the Middle East, and the great powers followed suit. Even when Ecevit was no longer prime minister his policies were still followed, despite much dissimilarity between him and his successor, Süleyman Demirel. But the efforts to improve Turkey's relations with the oil producers and the Third World at large turned out to be more difficult than expected and were fraught with disenchantment.

One major disappointment was Turkey's attempt to come closer to the nonaligned nations. In the summer of 1978, when Ecevit was again prime minister, Turkey tried to obtain the status of observer or guest at their summit—hoping, among other things, to get those nations to support its position on Cyprus. This attempt was an unmitigated failure. The Belgrade Conference, despite Ecevit's overtures, turned a cold shoulder to Turkey and expressed support for the Greek position on Cyprus. The problem, said a distinguished commentator, was that Turkey was trying to play both options at once.[16] Ecevit was wrong in supposing Turkey could belong to the developed Western world and the developing nonaligned world at the same time. Turkey had to decide whether it was a member of the "rich men's club" or the "poor men's union." Actually, the dilemma was rooted in the structure of the Turkish economy—an industry that could benefit from the existing world order (provided it emerged intact from the current crisis) on the one hand and an infrastructure still in need of "soft" credit and foreign technical know-how, which could perhaps become more readily available under the new economic order, on the other hand.

The oil producers, primarily Libya, encouraged Ecevit in his policies toward the Third World and the new economic order. Libya's leaders sought to persuade Turkey that the emergence of a new force in Africa and Asia would benefit the Turkish people. A joint statement issued by Turkey and Libya after signing their 8 January 1975 agreement supported the Third World in its struggle against colonialism, racism, and exploitation and affirmed the right of Third World nations "to develop and protect their natural resources," condemning any attempt to rob those nations of their total sovereignty over such resources.[17] Oddly, Turkey assumed this extreme posture just a few weeks after becoming a member of the group of countries that had founded the International Energy Agency (IEA). In February 1978, shortly after Ecevit had returned to office, the two nations decided to set up a joint committee charged with guiding Turkey-Libya cooperation in such a manner as to contribute to the creation of a new international economic order and thwart imperialism.

Demirel and the AP took a more guarded view of the Third World. They were angry with the oil producers for what they had done to Turkey. This attitude was not overtly reflected in official policy statements, but the AP never regarded the southeasterly direction as a real option for Turkish foreign policy. Demirel still believed in the West, viewing it as the only international option actually available to Turkey. Under his rule, Turkey continued to regard itself as a developing country and to support the creation of a new economic order, but that stemmed more from fear of possible retaliation by the oil producers than from any real belief in the ability of OPEC and the NIEO to put right the "historic injustices" Ecevit and his ministers habitually denounced. The established philosophy under Demirel was that in this modern world, where nations have become increasingly dependent on each other, the continued impoverishment of the Third World developing countries would create enormous problems for everyone. In the long run, even the developed nations would find that a new, more equitable international economic order would coincide with their own interests.[18]

The AP's attitude toward the Third World was thus strictly opportunistic (in that, it was little different from the attitude of many Western governments), in marked contrast to the posture assumed by Ecevit's CHP. The CHP leadership criticized Demirel's position as simplistic and lacking in depth. Former foreign minister Güneş described it as "[as] appropriate to the world as it was twenty years ago." Ecevit was furious with Demirel's refusal to regard the southeast as an option in the first place while sticking to a particular view of the world in which "the answers to all [Turkey's] problems emerge automatically."[19]

Another CHP spokesman, Hikmet Çetin, claimed Demirel's Western-oriented policies ran against the grain of Turkish interests.[20] According to CHP leaders, now in opposition, Demirel and his foreign minister, İhsan Çağlayangil, while appearing to support the NIEO, were actually seeking to hitch Turkey's foreign policy to the Western option alone: "Under Turkey's unique circumstances, there is a contradiction when one remains loyal to the traditional orientation while talking about new foreign policy directions. We shall eliminate this contradiction once we are back in office."[21]

A few years later, with Ecevit back in office, Turkey hosted an international seminar on the NIEO in August 1978 and in October 1979 initiated a roundtable forum of ministers of industry from various developing countries, intended to promote the NIEO and increase industrial cooperation among developing countries. Furthermore, Turkey even suggested establishing OPEC-like international cartels for tea and cotton exporters.[22] Realistically, such ideas were far-fetched, yet they were indicative of Turkey's new posture.

Ecevit's ideas on economic integration in the Middle East and a southeastern option lay behind the series of barter agreements with various oil producers in mid-1978. His ideological precepts, which inspired Turkey's political and economic policies in the international arena, were also reflected in his address at a meeting of Turkey's ambassadors to Middle Eastern capitals:

> Turkey can make an important contribution to Middle East solidarity: once our Turkish agriculture, industry, manpower, and infrastructure were harmoniously integrated with the natural resources and capital of other Middle Eastern nations, there would be no economic obstacle we shall not be able to overcome, no objective we shall not be able to achieve. Guided by this belief, we have endeavored to increase cooperation between Turkey and Middle Eastern nations since our return to office.[23]

Thus during the entire period under discussion (1974–1981), one can find virtually no public criticism by a Turkish leader of the oil embargo or the hard-hitting oil price hikes; it is far easier to find statements supporting those moves. The CHP was driven by an emotional sense of solidarity with the Muslim producers and support for the NIEO: Ecevit truly believed Turkey could assist in the oil producers' development and effectively contribute to a just world order. The AP, however, was less than enthusiastic. When Demirel returned to government in 1979, the Western option again became the dominant one, and it remained so under Turgut Özal and subsequent governments.

Turkey and the Islamic Conference

The changing patterns of Turkey's relationships with other Muslim nations affected in the long run Turkish attitudes toward Islam as an international force. The development of Turkey's relations with the most important organization in the Muslim world, the Islamic Conference, illustrates this trend.

Until 1969, the officially secular Turkey consistently refused to become involved in any alliance based on Islam, a mainstay of the Atatürk legacy. Thus Ankara rejected both the idea of an Islamic alliance suggested by the Somali prime minister in 1965 and King Feisal's call for an Islamic alignment in early 1966. The opposition to Turkey's participation in international Islamic forums was led by İsmet

İnönü, the prime minister for many years and a leader of the Kemalist revolution. Other leaders also felt Turkey's joining an Islamic front could establish a pattern of creating religiously based blocs in the international arena, eventually resulting in an open conflict with the West.[24]

In late 1969 the question came to the fore again, this time with a vengeance. A few weeks after a Christian lunatic set fire to the al-Aqsa Mosque in Jerusalem, Morocco's King Hassan sent a personal letter to Turkey's president, Cevdet Sunay, inviting him to attend an Islamic summit convened by the king to discuss the fire and its aftermath. At the same time, the ambassadors of Saudi Arabia and Jordan in Ankara explained to Turkish officials the significance their governments attached to high-level Turkish participation.[25]

Turkish opinion on the question of participation was divided. The CHP, then in opposition, claimed that any Turkish presence at a conference based on religious solidarity was unconstitutional, citing an article in the constitution that determined that the affairs of the state should be conducted on a secular basis. Besides, the party argued, participation would constitute a significant and undesirable shift in Turkish foreign policy. İnönü, the party's leader, thought that by joining the summit Turkey would involve itself in internal Arab affairs, which would go against the grain of its traditional foreign policy.[26] Prime Minister Demirel, however, considered that participation would enhance Turkey's relations with the Muslim world and determined that Turkey would take part in the conference and would actively participate in its proceedings, "so long as this would be consistent with the national interest and foreign policy considerations."[27] To minimize opposition, it was decided that Turkey would be represented by its foreign minister rather than the president. Turkey therefore informed the organizers that because of previous obligations involving the elections campaign then under way, neither the president nor the prime minister would be able to attend.

The September 1969 Rabat Conference was the first in a series of Islamic gatherings at the heads of state or prime minister level in which Turkey took part (see Table 7.1). Until 1975 the level of Turkish representation was consistently one rank lower than that of member states; that is, the Turkish foreign minister would go to prime ministers' conventions, and the director-general of the Foreign Office would attend foreign ministers' meetings. Also, during the years 1969–1974 Turkey regularly tabled its qualifications to resolutions adopted at the various gatherings of the Islamic Conference. Those qualifications stated that

Table 7.1   Islamic Conference Gatherings and Turkish Level of Representation, 1969–1989

Type of Meeting	Place	Date	Turkey's Representation
1st summit	Rabat	September 1969	foreign minister
1st foreign ministers	Jeddah	March 1970	director-general (D-G) Foreign Office
2nd foreign ministers	Karachi	December 1970	D-G Foreign Office
3rd foreign ministers	Jeddah	February–March 1972	D-G Foreign Office
4th foreign ministers	Benghazi	March 1973	D-G Foreign Office
2nd summit	Lahore	February 1974	foreign minister
5th foreign ministers	Kuala Lumpur	June 1974	D-G Foreign Office
6th foreign ministers	Jeddah	July 1975	foreign minister
Emergency foreign ministers	Jeddah	November 1975	D-G Foreign Office
7th foreign ministers	İstanbul	May 1976	foreign minister
8th foreign ministers	Tripoli	May 1977	foreign minister
9th foreign ministers	Dakar	April 1978	foreign minister
10th foreign ministers	Fez	May 1979	foreign minister
Emergency foreign ministers	Islamabad	January 1980	foreign minister
11th foreign ministers	Islamabad	May 1980	foreign minister
Emergency foreign ministers	Amman	July 1980	D-G Foreign Office
Emergency foreign ministers	Fez	September 1980	foreign minister
3rd summit	Taif	January 1981	prime minister
12th foreign ministers	Baghdad	June 1981	foreign minister
13th foreign ministers	Niamey	August 1982	foreign minister
14th foreign ministers	Dacca	December 1983	foreign minister
4th summit	Casablanca	January 1984	president
15th foreign ministers	San'aa	December 1984	foreign minister
16th foreign ministers	Fez	January 1986	foreign minister
5th summit	Kuwait	January 1987	president
17th foreign ministers	Amman	March 1988	foreign minister
18th foreign ministers	Riyadh	March 1989	foreign minister

Turkey would implement such decisions only subject to its own constitution and foreign policy, as well as to UN principles. This effectively relieved Turkey of any obligation to implement any resolution, because its constitution ruled out the adoption of any resolution taken by a religiously based body. Also, the long-standing principles of Turkey's foreign policy (which included the notion that Turkey should maintain friendly relations with all nations of the world) did not coincide with resolutions that called for a "holy war" against Israel or severance of relations with that country. This specifically meant Turkey, which at the time was the only Muslim nation (besides Iran) that had diplomatic relations with Israel.

The Lahore summit (22–24 February 1974) was the first gathering of the Islamic Conference after the outbreak of the 1973 energy crisis. As part of his preparations for the summit, Conference Secretary-General Hassan al-Tohami went to Ankara to find out from Foreign Minister Haluk Bayülken what Turkey thought of three draft resolutions that might have proved problematic from its point of view: a call for a jihad (holy war) against Israel, a demand that member nations sever relations with Israel, and a call for all Muslim nations to adopt the Arabic alphabet (which Turkey had abolished, replacing it with a Latin one during the Kemalist revolution). Bayülken rejected all three notions, which the Turkish press also saw as outrageous.[28]

Nevertheless, oil was placed high on the agenda of the Lahore Convention, so Turkey had great expectations. It was decided that the new foreign minister, Turan Güneş, would be accompanied by several senior officials who, it was hoped, would be able to conduct bilateral talks with leaders of the major oil producers on cheaper oil for Turkey.[29] That team was thus the strongest Turkey had ever sent to an Islamic gathering.

Still, the resolutions adopted by the Lahore Convention on aid to Muslim nations hurt by the oil price hikes were rather vague. It was resolved that problems created by the oil crisis, as well as the entire issue of economic relations among Muslim nations, ought to be settled "as part of the overall understanding." This noncommittal sop made Güneş and his associates pessimistic about the possibility of receiving any real assistance from the oil producers.

Even though the Turkish delegation had gone to Lahore hoping for aid from the oil states, it adamantly rejected—during the convention and later on—the resolution that called on it to sever relations with Israel; instead, Turkey presented its usual qualifications. Returning to Ankara, Güneş flatly announced that Turkey had no intention of obeying the resolution. Officials added that such a notion was inconceivable while Middle East peace talks were going on, both in the Sinai and in Geneva. The Lahore resolution was dubbed "wishful thinking."[30]

Ankara's unequivocal refusal made the Muslim nations, particularly Saudi Arabia, turn their efforts to persuading Turkey to become a full member of the Islamic Conference (still called the Union of Islamic States). This request was put directly to Deputy Prime Minister Erbakan during his visit to Riyadh in April 1974. The Saudis argued that Turkey's official secularity need not be an obstacle to joining. King Feisal even gave Erbakan four examples of members whose constitutions included a "secularity clause," including Lebanon, which joined the conference even though a large part of its population was not Muslim. Besides such arguments, it was also reported that the Saudis made the issue of credit for Turkey contingent on its joining, because only by doing so could Turkey become eligible to receive loans from the Islamic Development Bank.[31]

Still, Turkey officially rejected the Saudi appeal, and Erbakan's announcement during his visit that he personally favored having Turkey become a full member created a political storm. Senator Kamran İnan called Erbakan's attempt to make Turkey a full member of the Islamic Conference "a rape of [the] Turkish constitution." He continued, "The values developed by Turkey for 50 years as a republic cannot be exchanged for all the oil in the world."[32] The firm resistance (not only by the CHP but also by many in the AP) to Erbakan's pro-Islamic proclivities laid to rest for a time the question of Turkey's membership in the Islamic Conference—but only for a while.

In view of the increased power garnered by the oil producers, the need to drum up Muslim support on Cyprus, and the U.S. arms embargo imposed on it, Turkey soon significantly changed its attitude. In July 1975 it was decided that Turkey's representative at the Islamic foreign ministers' convention in Jeddah would be Foreign Minister İhsan Sabri Çağlayangil. This was unprecedented, because it put Turkey's level of representation on a par with that of the other participants. Moreover, Çağlayangil asked to host the next annual foreign ministers' gathering in Istanbul. His delegation also somehow neglected to table its traditional qualifications, even though the anti-Israel resolutions were particularly tough.[33] This was also the first time the Cypriot Turkish leader, Rauf Denktaş, was allowed to address the convention.

The most important issue facing Ankara, however, was whether to ratify the charter of the Islamic Conference—a move that would have amounted to assuming full membership.[34] The Turkish government also changed its position on this matter during 1975–1976: Ecevit's strong position in 1974 had evaporated by 1976. On 10 May 1976, just before the Islamic foreign ministers were scheduled to convene in Istanbul, the director-general of Turkey's Foreign Office, Şükrü Elekdağ, announced that the government had decided to ratify the charter of the Union of Islamic States and to join the organization as a full member. He emphasized that Turkey would maintain its policy of tabling qualifications to any resolution that did not coincide with its constitution. Two months later Prime Minister Demirel confirmed in a press conference that Turkey's government had decided to ratify the charter, "subject to constraints imposed by our constitution." He added that a bill to that effect had already been brought up in parliament.

The government launched a campaign to explain its new policy to the public and succeeded in winning widespread support from all major political parties. The only senior official who still had reservations was President Fahri Korutürk, who elected to tour the eastern part of the country rather than preside over the convention's opening ceremony in Istanbul. Çağlayangil explained that the Islamic Conference was a political rather than a religious organization; hence joining it did not contradict the constitution.[35] He said that even prior to his government's decision, member states had treated Turkey as a full member, as evidenced by the fact that the convention was taking place in Istanbul. Therefore the decision only meant changing Turkey's status from a de facto to a de jure member. He explained further that the international situation created by the Cyprus War had made the government decide to join the conference, as Turkey had failed to find support in the West for its moves in Cyprus and the U.S. arms embargo had led public opinion to support a neutralistic foreign policy. In conclusion, the government saw no reason to object to full Turkish membership in the organization.

A Foreign Office official, Hamit Batu, had an additional explanation.[36] Over the centuries, common beliefs and shared values, precepts, traditions, and hopes had forged strong bonds of fraternity among the Muslim nations. The purpose of the Islamic Conference was to foster awareness of common traditions and heritage and to use that awareness to solve contemporary problems; thus it was inconceivable that Turkey, as a leading nation in Islamic history, would stay outside such an organization. The question of whether Turkey's official secularity precluded such a move was irrelevant. Turkey did separate religious and state affairs, but that separation was not intended to prejudice any religious ties Turkey may have had or to affect the "natural ties" binding the Turks to their coreligionists in other countries.

At the foreign ministers' convention in Istanbul, horse trading between Turkey and conference members was intense. As a reward for the Turkish government's forthcoming decisions to ratify the Islamic Conference charter and to allow the Palestine Liberation Organization (PLO) to open an office in Ankara (the decision regarding the PLO had been announced one day before the convention began) and for Turkey's support of the anti-Israel resolutions to be adopted during the convention,[37] the organization agreed to invite a representative of the Turkish Cypriot community to all its future meetings. The convention also called for equal rights for both Cypriot communities and adopted a series of resolutions on economic affairs that were in line with Turkish interests: Turkey's offer to host a meeting of Islamic ministers of economic affairs, industry, and technology was accepted; a center for economic and social studies of Muslim nations was to be established in Ankara; a permanent social and economic committee was to be created as an organ of the conference; and a center for the study of Islamic history, art, and culture was to be established in Istanbul.

Yet despite all the government's official statements and endeavors to bring about ratification of the Islamic Conference charter, parliament refused to do so. Turkey has remained only a de facto member.

At the eighth foreign ministers' convention (Tripoli, May 1977) and during the following months, Turkey kept up its efforts to convert political cooperation within the Islamic Conference to economic benefits. A draft proposal produced by a committee headed by Turkey, calling for economic, commercial, and technological cooperation, was adopted; and in October 1977 the convention of Islamic ministers of economic affairs, industry, and technology (agreed upon at the Istanbul convention) took place in Ankara. But Turkey's struggle to persuade the Muslim world of the need to provide economic assistance to needy Muslim nations failed miserably. The Ankara Convention, masterminded by Deputy Prime Minister Erbakan, was stillborn. Only one ranking minister (Libya's minister of industry) arrived at "Erbakan's Convention"; all other member states sent low-ranking officials. Erbakan barely managed to save some prestige when the president of the Islamic Development Bank announced a $10 million loan to the İskenderun steelworks.

An emergency meeting of Islamic foreign ministers convened in Islamabad in January 1980 put Turkey in a particularly vulnerable position. Whereas previous Islamic gatherings had dealt mainly with the Arab-Israeli conflict, the Islamabad meeting had at the top of its agenda such issues as the Soviet invasion of Afghanistan, the U.S. embargo against Iran, and the Muslim nations' relations with Egypt following its peace accord with Israel. A strong anti-Soviet resolution was tabled, at a time when Turkey was eager to maintain good relations with the USSR. This time Turkey was unable to hide its reservations behind a polite curtain of qualifications made after the event, as it had done in the late 1970s with regard to resolutions on the Middle East conflict. It now had to make its reservations plain and public. Turkey openly opposed a Saudi call to declare a jihad on the Soviet Union because of Afghanistan, describing it as "cutting against the grain of modern world circumstances"; Turkey also differed with most conference members on such issues as sanctions against Iran and participation in the Moscow Olympic Games. The need to remain on friendly terms with the Soviets made Turkey's relations with Iraq, Libya, and Saudi Arabia rather awkward. Still, Iraq and Libya did not retaliate in terms of oil supplies, much to Turkey's relief.

The Generals and the Conference

Between November 1980 and January 1981, against the background of the Iran-Iraq War and the oil crisis it produced, the Islamic Conference went into high gear. A third summit was convened in Taif, Saudi Arabia. Turkey was heavily involved in the preparations for the convention, striving as ever to have it give more weight to economic issues. In particular, Ankara sought to have previous economic resolutions implemented. In addition, the Turks worked out a joint development strategy for the Islamic nations that was called, after various modifications, "the Ankara plan." It called for increased economic assistance to fifteen underdeveloped Muslim nations, cooperation on energy, and discussions on an Islamic common market. Turkish lobbying efforts produced a draft resolution calling on OPEC nations to give Muslim nations priority in oil supplies. The resolution was eventually accepted but in a vague, unbinding manner, because it left the final decision to each producer government.

Opened on 24 January 1981, the Taif Convention was attended by Prime Minister Bülent Ulusu. Participants were not overly fazed by the fact that Turkey was represented by an appointee of a military junta— especially since the Turkish generals thought highly of the Islamic Conference. Addressing the convention on 27 January, Ulusu said the Turkish people expected the Taif summit to become a turning point in efforts to increase economic cooperation in the Muslim world, proving that it was united "in faith, in objectives, and in hopes for peace." Summit resolutions on cultural affairs (setting up cultural centers, universities, and an Islamic technological center) would enhance "grassroots solidarity" among member states. The Islamic Conference could potentially contribute not only to the well-being of its members but also to world peace and stability. Ulusu congratulated the summit for adopting the Ankara plan and expressed his gratification that "political solidarity among the Muslim nations is now finding an expression in increased economic cooperation."[38]

Fundamentally, there were two reasons for Turkey's growing involvement with Islamic affairs: the need to move economically and politically closer to the oil-producing countries and Saudi Arabia, first and foremost, and the need to rally widespread support for Turkey's position on Cyprus. It is difficult to tell how much Turkey's activities within the Islamic Conference actually affected its relations with such oil producers as Iraq or Libya, with which it had multifaceted relations anyway. But regarding Saudi Arabia and Kuwait, with which Turkey had little in common in the area of bilateral relations, it is noteworthy that the shift in Turkey's position on the Islamic Conference—from nearly complete detachment in 1975 to intensive involvement by 1981—barely affected its relationships with those nations. Thus, although Turkey had received some aid from the Saudi-controlled Islamic Development Bank, Riyadh was still adamant in its refusal to sell oil directly to Turkey, as noted earlier. This was surprising in view of the importance Saudi Arabia had attached to Turkish involvement in the Islamic Conference. Relations between Ankara and Riyadh did begin to improve in 1981, but that was mainly the result of changing international economic circumstances.

Turkey has paid a considerable price for the Islamization of its foreign policy. According to former prime minister Bülent Ecevit, that move has undermined the secular foundations of the Turkish republic and jeopardized the national identity it has worked to create since the Atatürk revolution. Having begun to efface its unique identity, he said, "Turkey was being pulled and pushed into a growing identification with the anachronistic world of Islamic traditionalism."[39]

Since 1984 Turkey has been represented at the various Islamic conventions at the nominal level. Thus President Kenan Evren attended the fifth summit (Kuwait, January 1987). Since the mid-1980s Turkey has added to its usual efforts to gain support for Cypriot Turks and has increased its diplomatic maneuvers on behalf of the Turkish minority in Bulgaria. And indeed, both the Kuwait summit and the next foreign ministers' convention (Amman, March 1988) referred to that issue in their resolutions. Still, Ankara was bitterly disappointed at the Muslim states' attitude toward both Cypriot and Bulgarian Turks, especially once Bulgaria started its policy of deportation. Despite intensive lobbying, no real results were achieved on either issue, and a shadow was cast over Turkey's relations with many Muslim states.

Islamic Reaction in Turkey

As early as the 1960s some external radical Islamic groups (especially Saudi and Jordanian ones) were involved in Turkey's religious and social life.[40] These activities were masterminded by a Palestinian named Ahmad al-'Ali, a member of the Beirut-centered Hizb al-Tahrir al-Islami (Islamic Liberation Party). Al-'Ali's group organized public rallies in Ankara and Konya and opened branch offices in those cities. In August 1967, when Hizb al-Tahrir began distributing leaflets of a political character, some of its leaders were arrested. It then transpired that Jordanian students at Ankara University were involved in these activities.

During the 1980s foreign nations, particularly Iran and Saudi Arabia, became increasingly involved in Islamic activities in Turkey. Ankara was particularly disturbed by Iranian radio broadcasts calling for the establishment of an Islamic state as the solution to Turkey's domestic problems. There was also much criticism of Saudi involvement in mosque building and religious studies, both in Turkey proper and among Turkish workers abroad.

In a very interesting article, Ali Karaosmanoğlu analyzed the overall reasons for this multifaceted involvement—Iranian propaganda broadcasts, Libyan statements on domestic Turkish problems, external financing of Islamic activities inside Turkey—which he regarded as dangerous to Turkish integrity.[41] As he put it, the sociopolitical and cultural development of most Muslim nations in the Middle East had not yet created the requisite conditions for a rational foreign policy free of religious constraints. At the same time, nationalism in those countries still included a strong anti-Western component, and its revolutionary style (Qaddafi, Khomeini, Assad) was responsible for much of the endemic instability in the region. In Turkey, however, Kemalist reforms (secularization, Westernization) had created the necessary conditions for the realism on which foreign policy was based. Turkish policies in the 1980s had set up objectives according to capabilities and examined ideas in a pragmatic, down-to-earth manner. Free of religious considerations, that policy could be called bureaucratic-pragmatic. Still, Turkish decisionmakers were not sufficiently aware of the gap between the unrealistic policies of other nations in the region and their own realism. Karaosmanoğlu thought such lack of awareness could open the door to Islamic subversion, which Turkey could find difficult to counter. By portraying Islamic nations as allies and declaring its unconditional support for the Arab position in the Arab-Israeli conflict, the Ankara government was unwittingly setting the stage for activities that might become subversive—sometimes without any checks and balances.

The most notorious case of outside intervention, called "the Rabita affair," shook Turkey in the early 1980s. Rabitat al-'Alam al-Islami (the World Islamic Union) was a Saudi organization whose stated aim was to establish a pan-Islamic federation based on Islamic law. Drafted in Mecca in 1963, the organization's constitution charged it with ensuring that Muslim states followed the laws of the Quran. In 1987 the Turkish press published a story (that had been known in diplomatic circles in Ankara for some time) to the effect that as far back as 1981 the junta-appointed Ulusu government had given Rabita permission to pay the salaries of Turkish religious functionaries operating among Turkish workers in Europe. Furthermore, the newspapers claimed, the organization was also allowed to provide financial support to religious organizations and projects in Turkey proper. It seems the organization had access to the highest officeholders in the country. Those relations had been cemented in 1980 and 1981—when Turkey was busy courting Saudi Arabia for oil and loans.[42]

According to various sources, during the 1980s the Saudis financed Arabic-language study programs, gave financial support to religious orders in Turkey, made contacts with religious research institutes and projects, and helped a number of Islamic organizations, newspapers, and publications. Rabita money was even involved in the construction of a mosque inside the parliament buildings and in the establishment of an Islamic center at Ankara University (a project proposed by Korkut Özal, the prime minister's brother). The Rabita affair proved, according to left-wing circles, that Saudi Islamic activities in Turkey went way beyond normal economic involvement and that the Saudis received benefits not normally accorded foreign investors.

In March 1987 President Evren reacted to these revelations by owning that Rabita had indeed been granted permission to operate in these spheres, which he dubbed a mistake in retrospect. Revealingly, the president explained the decision in reference to Turkey's difficult economic situation in the early 1980s.[43] He said the state should have undertaken the financing of those projects, but it was unable to do so at the time. This was one of the first open admissions that during the early 1980s Turkey had been forced to pay for its improved economic relations with Saudi Arabia with cultural and religious concessions.

Curiously, President Evren was almost alone among Turkish officials in responding to the Rabita affair accusations. Other officials remained silent. This was unusual because Turkey's government had always been quick to react whenever it perceived outside interference in its internal affairs. Thus in January 1987 Ankara issued an official protest to Iran's president, Hussein Mussavi, when Iran weighed in on the ongoing public debate regarding females wearing veils on university campuses. Radio Ankara also called on Turkey's neighbors to refrain from interfering in its internal affairs. On another occasion Iran's ambassador was summoned to the Foreign Office and rebuked for his participation, alongside Necmettin Erbakan, in an anti-Israel demonstration organized by the Refah Partisi (Welfare Party), at which Turkey's official policy of secularization was also denounced.

Iran's involvement in the debate on female students at Turkey's universities wearing veils was particularly outrageous in Turkish eyes. The ban on people who were not recognized religious officeholders wearing articles of clothing that constitute religious symbols had been imposed as far back as 1925. Mustafa Kemal Atatürk himself attached supreme importance to this regulation, as he explained in 1927: "It was necessary to abolish the fez, which sat on the heads of our nation as an emblem of ignorance, negligence, fanaticism and hatred of progress and civilization."[44] On another occasion he expressed a similarly strong view on the veil:

> In some places I have seen women who put a piece of cloth or a towel or something like it over their heads to hide their faces, and who turn their backs or huddle themselves on the ground when a man passed by. What are the meaning and the sense of this behavior? Gentlemen, can the mothers and daughters of a civilized nation adopt this strange manner, this barbarous posture? It is a spectacle that makes the nation an object of ridicule. It must be remedied at once.[45]

And indeed, conservative Islamic circles reacted furiously to the ban as soon as it was imposed. Egypt's highest religious authorities announced in 1926 that anyone who replaced the fez with Western headgear was a heretic.

Early in 1987 a prolonged public debate began when the Higher Education Council issued an internal ordinance banning veils on campuses. A leader of the Anavatan Partisi (Motherland Party), Mehmet Keçeciler, challenged the ban and asked the Ministry for Religious Affairs to intervene. The left wing reacted furiously. President Evren joined in the debate on the side of secularity. But the religious circles enjoyed widespread political support (including, significantly, that of Prime Minister Özal), in addition to much encouragement from Iran. They argued that the ban infringed on personal freedom, another constitutional principle. Thus the debate over this apparently trivial matter became symbolic of the struggle between Kemalism and fundamentalism, and emotions ran high.

In 1985 former prime minister Ecevit published a series of articles on the issue in support of secularity and against Muslim interference. His attitude toward the Muslim world seemed to have undergone a significant change since he left office. At the time, when the energy crisis had been his chief concern, he spoke softly; now he wielded a big stick: "The Turkish system and the example it has set have constituted the greatest threat to the anachronistic, theocratic, and dictatorial regimes in most Muslim states. Turkey's way was a thorn in the sides of the forces of oligarchy in several Muslim societies."[46] Those forces, he said, felt a need to attack Turkish secularity and undermine the accompanying processes of democratization and modernization—even to destroy secularity, if they could, through propaganda and subversion. One of the main tools of the "Muslim oligarchies" endeavoring to change Turkey's religious and cultural character and undermine its national identity, he believed, was the Islamic Conference and its various organs.

Islamic Reaction and Foreign Policy

The Turks have never questioned their Muslim affinity. Prime Minister Özal said, when on a pilgrimage to Mecca in July 1988, "Turkey as a state is secular, but not I; I am a Muslim." Even the staunchest supporters of Kemalist reforms understood the difference between a secular and an atheist state. They attacked the religious establishment, not Islam as a religion. The question has never related to anyone's personal beliefs; rather, it has concerned the role of Islam in public life, particularly politics and the justice system.

Scholars dealing with the problems of Islam in Turkey commonly accept that in the foreseeable future Islamic reaction constitutes no existential threat to the fundamentals of secular republicanism there. One scholar who has written much on this subject is Doğu Ergil, who gives three principal reasons for this belief:[47]

1. Overall, the activities of Islamic circles are no longer clandestine. Hence they can be closely monitored by the authorities.
2. Turkey's Western orientation is widely regarded as a success. It has borne fruit and enjoys widespread support. Most people believe in democracy and object to the establishment of political parties on a religious basis.
3. The Turkish left wing is secular and unlikely to join in a coalition with religious circles in an attempt to make fundamental changes in Turkey's political system.

Ergil's optimistic conclusions (which refrain from mentioning the usual safety valve, a military coup d'état) are shared by few among the Turkish left wing, where the view prevails that Islamic reactionism is still a threat to Turkey's democratic institutions and secular order. One of the chief supporters of this view, as mentioned earlier, was former president Evren. The view has also been widely held by chiefs of the Turkish military, particularly during the second half of the 1990s. And even scholars who did not foresee an Islamic fundamentalist takeover still considered it possible that militant religious groups might destabilize and hurt democracy.[48]

Thus the scope and objectives of Islamic reaction in Turkey must be considered carefully. One of the reasons for this concern is political rather than religious in nature. A strong tendency to get closer to other Muslim nations has become a conspicuous characteristic of the Turkish religious camp, and past experience (particularly during the 1970s) has shown that it does not take a major change in the structure of the regime itself to bring about policy changes in that direction. Experience has also shown that a foreign policy of rapprochement with the Muslim world can have far-reaching implications on the status of Islam in Turkey.

The struggle between Western democracy and the Muslim world over Turkey's destiny is largely hidden and is far from over. Any adverse development in Turkey's relations with the West—harsh economic demands by the European Union, for instance, or U.S. demands on defense—would strengthen the anti-Western camp and the Islamic element; the reverse is also true. At the time of writing, as the European Union was expanding eastward to include former Warsaw Pact members, NATO was increasingly losing its erstwhile importance. In consequence, the importance of Turkey to Europe could decline, strengthening Islamic proclivities inside Turkey. Indeed, several Turkish leaders, including former prime minister Mesut Yılmaz, have indicated that if the European Union rebuffs Turkey, "we shall have to look for our fortunes within different communities of nations."[49] However, the recent election of President George W. Bush in the United States might strengthen the U.S.-Turkish link, especially if the Republican administration will revive its harsh anti-Iraqi, and maybe even anti-Iranian, policy and need Turkey as a buffer.

Moreover, the role of Islam in Turkey cannot be analyzed in isolation from other domestic problems, first and foremost of which is Kurdish separatism. This is an ongoing problem for the regime, and some consider it a real threat to the integrity of the state itself. In the past, tensions between Sunni and 'Alawi Muslims in Turkey were reduced when secularism was made supreme. It may turn out, however, that tensions between Turks and Kurds are reduced by promoting Islam, the most significant spiritual link between Turkey's Kurds and the rest of the nation. Thus if Kurdish cultural demands grow stronger, Turkey may find it advisable to promote Islam as a common uniting bond.

Notes

1. Sezai Diblan, president of the Turkish Chamber of Commerce, said in late 1977: "If the same determination displayed with regard to oil would also be displayed with regard to wheat, tobacco, and cotton, Muslim nations would enjoy higher incomes." *Pulse*, 19 October 1977.

2. *Istanbul Bayram*, 6 January 1974. One month later, on 5 February 1974, *Pulse* wrote in an editorial: "In view of the fact that the prime minister praised the perception of nationalism among the Arab shaikhs, and his deputy described the energy crisis as a once-in-a-lifetime opportunity, it should not be too difficult to guess where Turkey chose to place its bet on the outcome of the struggle between oil exporters and consumers."

3. Ecevit's address was published in Turkish newspapers on 8 August 1978.

4. *Cumhuriyet*, 17 February 1974.

5. Interview with the author, Ankara, 16 May 1986. Baykal also noted that Turkey's leaders realized all too well that verbal objections to the oil price increases would have had no impact in the international arena.

6. Interview with the author, Ankara, 20 May 1985.

7. A leading MSP member told *Briefing* on 22 August 1975: "After a long period of economic exploitation by the West, Turkey now needs the friendship of nations with whom it has common religious and spiritual grounds. This means the Arab world."

8. *Pulse,* 4 July 1974.

9. *Cumhuriyet,* 25 May 1976.

10. *Pulse,* 24 February 1978.

11. *Pulse,* 8 August 1978.

12. *Milliyet,* 5 January 1976.

13. *Pulse,* 31 July 1979.

14. At the time, this concept was the subject of intense discussions in various international forums. Basically, the NIEO meant changing international trade and investment patterns so as to make developing countries equal partners with the developed ones rather than suppliers of cheap raw materials and consumers of expensive finished products.

15. Yılmaz Altuğ, "Turkish Foreign Policy Vis-à-Vis Recent World Developments," *Annals of the Faculty of Law,* University of Istanbul (1976), pp. 71–84.

16. Martin Woollacott, "Turkey Eyes the Third World Again," *Guardian,* 18 August 1978.

17. *Pulse,* 9 January 1975.

18. Hamit Batu, "New Developments in Turkish Foreign Policy," *Foreign Policy* (Ankara) 5, no. 4 (1976), pp. 5–17.

19. *Pulse,* 19 June 1975.

20. *Cumhuriyet,* 24 February 1980.

21. Ecevit in an interview in Yugoslavia, quoted in *Pulse,* 25 May 1975.

22. Ecevit brought up the idea of a tea cartel during a visit by the president of Bangladesh to Turkey in October 1978. In March 1980, during a conference of seventeen cotton-growing nations in İzmir, Turkey's representative suggested a cotton cartel.

23. *Pulse,* 30 April 1979. Similar reports appeared in various other Turkish newspapers.

24. Asked for his views on the subject in 1965, İnönü replied: "I wonder how we would have felt about it, if the Christian nations [had] decided to sign an international Christian treaty." See Kemal Karpat, *Turkish Foreign Policy in Transition* (Leiden: E. J. Brill, 1975), p. 175.

25. Omer Kürkçöğlu, *Turkiyènin Arap Orta Dogusùna Karsi Politilasi* (Ankara, 1972), pp. 167–169; *Cumhuriyet,* 14 September 1969. Not all Muslim nations were keen about Turkish participation in the summit. Egypt, for instance, was vehemently opposed because of Turkey's ties with Israel.

26. Ibid., pp. 167–168.

27. *Disileni Bakanligi Belletini,* no. 60 (September 1969): 12–20.

28. *Pulse* (3 January 1974) claimed in an editorial that accepting Tohami's proposal would have meant "mortgaging Turkey's foreign policy."

29. It was later revealed during a Senate debate on an unrelated issue that at the time several Muslim nations, including Saudi Arabia, had exerted pressure on Turkey to put Deputy Prime Minister Erbakan, rather than Foreign Minister Güneş, at the head of its delegation.

30. *Milliyet,* 26 February 1974.

31. *Hürriyet,* 6 May 1974.

32. Senate debate regarding prime minister's budget, May 1974.

33. The letter of qualification was left with the Turkish ambassador in Jeddah, who chose to present it only after it had been decided that Istanbul would host the next convention. It also seems the qualifications were tabled only after the United States had strongly protested against the failure to do so during the conference. *Pulse,* 20 November 1975.

34. Adopted at the third foreign ministers' conference in Jeddah, February–March 1972, the charter included calls for solidarity and cooperation among Islamic nations in the economic, social, cultural, and scientific spheres, as well as political articles on the struggle against racial discrimination and the various forms of colonialism and the need to assist the Palestinian people.

35. *Milliyet,* 11 May 1976.

36. Batu, "New Developments."

37. The Istanbul Convention resolved to call for Israel's expulsion from the United Nations and other international organizations and declared Zionism "a colonialist, racist, and imperialist doctrine, posing a direct threat to international security." Turkey voted in favor of the resolution, and it is not clear whether it added its traditional qualifications. For the full text, see "Declaration of the Islamic Conference, 1976," *Disisleri Bakanligi, Belleteni,* June 1976, pp. 1–10.

38. *Pulse,* 30 January 1981. Turkish public opinion accepted with remarkable equanimity Turkey's new involvement in Islamic cultural affairs, bearing in mind the severe criticism Erbakan had received for his cultural agreement with Saudi Arabia just seven years earlier.

39. Bülent Ecevit, "The Interdependence of Turkey's Internal and External Policies," *Turkish Daily News*, 21–25 May 1985.

40. See J. Landau, *Radical Politics in Modern Turkey* (Leiden: E. J. Brill, 1974).

41. Ali Karaosmanoğlu, "Turkey's Discreet Foreign Policy Between Western Europe and the Middle East," in *Middle East, Turkey, and the Atlantic Alliance* (Ankara: Foreign Policy Institute, September 1987), pp. 90–93.

42. See Uğur Mumcu, *Rabita* (Ankara: Tekin Yayinevi, 1987).

43. *Briefing,* 30 March 1987.

44. Bernard Lewis, *The Emergence of Modern Turkey* (Oxford: Oxford University Press, 1968), p. 268.

45. Ibid., p. 271.

46. Bülent Ecevit, "The Interdependence of Turkey's Internal and External Policies," *Turkish Daily News* (part 2), 22 May 1985.

47. Doğu Ergil, *Secularism in Turkey* (Ankara: Foreign Policy Institute, 1989), pp. 76–78.

48. Ali Karaosmanoğlu, "Islam and Foreign Policy: A Turkish Perspective," *Foreign Policy* (Ankara) 12, no. 1–2 (June 1985).

49. Quoted in *Neue Züriche Zeitung,* August 1989.

8

POLITICS, OIL, AND ISLAM: RELATIONS WITH IRAQ, IRAN, AND LIBYA

The comprehensive set of agreements on energy and the economy between Turkey and Iraq had a political rationale in addition to the economic one. In the past, political relations between Ankara and Baghdad had developed to the point of signing a military alliance pact in 1955. The revolution in Iraq in 1958 eliminated the most important element in that alliance—attachment to the West—and it crumbled. But in the mid-1960s the two nations began inching closer to each other once again, as Ankara's foreign policy started to change from a staunchly pro-Western stance to a more diversified one. Besides, it was in the interest of both nations to cooperate on the Kurdish problem: Iraq had to deal with Kurdish national aspirations within its territory, and Turkey was facing repeated Kurdish attacks along its eastern border.

Iraqi Stability

Political relations between Turkey and Iraq were stable during the 1960s, and that made Iraq Turkey's preferred oil supplier—all the more so since Baghdad was showing interest in building an oil pipeline to the Mediterranean through Turkish territory. Such massive, long-term dependence on Iraq was not seen as a political liability for Turkey at the time, because Turkey's leaders seem to have believed relations between the two nations were stable. The economic agreement signed between them in August 1973 was regarded by Ankara as an important and promising achievement, in contrast to Iran's refusal to enter a similar agreement with Turkey. Iraq's political support during the Cyprus War in 1974 added another dimension to these multifaceted relationships.

There were three main reasons for Iraq to maintain good relations with Turkey during the 1970s and 1980s: (1) the Kurdish problem, with its ramifications for security along the Turkish-Iraqi border; (2) the allocation of Tigris and Euphrates water; and (3) Iraq's dependence on Turkey for land and sea transport routes.

The national aspirations of the Kurdish minority in Iraq had always been a source of grave concern for Baghdad, and it needed Turkish cooperation to curb Kurdish activities in the border area. Turkey was interested in thwarting similar aspirations by its own Kurdish minority and in maintaining peace and quiet in its southeastern provinces, but Iraq's problems were more acute. Iraqi Kurds lived mainly in the Kirkuk and Mosul areas, where more than half of the country's oil and gas was produced. Political unrest in those sensitive areas could have seriously hurt the Iraqi economy. When the Iraq-Turkey oil pipeline was built, crossing Kurdish areas, Ankara's political cooperation was needed even more. With the added problem of competition between Iraq and Iran over Turkish cooperation in their respective efforts to contain various elements within the Kurdish national movement, it is easy to see why the Kurdish issue put Ankara in a sound bargaining position vis-à-vis Baghdad.

One of the most difficult aspects of the Kurdish problem for Turkey has been the relentless violence against Turkish citizens in the border area. During the 1980s and 1990s the Turkish army's ongoing anti-terrorist efforts included dozens of military operations against Kurdish separatists inside Iraqi territory, mostly with Iraqi consent and collaboration. The first such action took place in May 1983, conducted by ground forces only; in the second (August 1986), in retaliation for the killing of twelve Turkish soldiers, the air force also took part. The Turkish armed forces attacked bases of the Parti-ye Kerkaran-î Kürdistan (PKK, Kurdistan Workers' Party), raising strong protests from Libya and Iran. Such operations inside Iraqi territory became almost routine in the 1990s.

The establishment of an independent, irredentist Kurdish state in northern Iraq would have had serious implications for the national aspirations of Turkey's own Kurdish minority, which constitutes one-fifth of the entire population. The national and cultural aspirations of that

minority have never found meaningful expression in the Turkish political system, but its extra-parliamentary presence has been considerable.
Since the mid-1980s one major political party, the Social Democrats
(then the second-largest party in parliament), has taken a position that
can be regarded under the circumstances as sympathetic toward the
Kurds. For instance, it criticized the 1986 military operation in parliament. Some of its parliamentarians even participated in an international
conference on Kurdish rights in Paris in November 1989, raising a
storm in Turkish politics.

An equally important issue in the bilateral relations between Turkey
and Iraq, at least as seen from Baghdad, has been water. More than 80
percent of Iraq's water resources flow into the country from the outside.
Saddam Hussein has often called the headwaters of the Tigris and
Euphrates, in Turkey's eastern mountains, the "vital arteries" of the
Iraqi economy. Since 1960 Turkey, Iraq, and Syria have been quarreling
over Euphrates water, a total annual average flow of 32 billion cubic
meters. Both Turkey and Syria have begun to develop the parts of the
river flowing through their respective territories, ignoring Iraq's
demand first to reach a binding agreement on allocation. Of particular
concern to Iraq have been Turkey's development plans for its southeastern provinces, which included twenty-one dams and seventeen hydroelectric power stations along both the Euphrates and the Tigris. Many of
those dams, including the Atatürk Dam—the ninth-largest in the
world—have been completed.[1]

In early 1990, when Turkey began filling the Atatürk Dam reservoir, Iraqi water supplies were badly damaged. Repeated protests to
Ankara by visiting Iraqi officials (including, intriguingly, the minister
of oil in January 1990) proved futile. The minister of oil even brought a
personal letter from Hussein to Turkey's president, which the Turkish
press interpreted as a hint of possible retaliation in terms of oil, but Iraq
took great pains to deny such speculation.[2]

Iraq made every conceivable effort to frustrate Turkey's development plans in the southeast, including lobbying the World Bank and
other oil producers. This had the effect of reducing the amount of external financing available for those projects, and the World Bank demanded that Turkey reach a prior agreement with Iraq and Syria on the allocation of Euphrates water. But Turkey has been persistent in pushing its
development plans ahead, albeit at a slower pace. At the time of writing,
no solution to this problem could be seen and the plans are progressing
as a Turkish national effort.

A third factor in Iraq's dependence on Turkey has involved the logistics of land transportation to and from Iraq, including the flow of oil. In view of the political-military situation, the pipelines from Iraq to ports in Syria or Lebanon have always had uncertainty attached to them. Syria stopped the flow of oil in the Iraqi pipeline cutting through its territory between December 1966 and March 1967, nationalized its part of the pipeline in 1972, and doubled passage fees in 1973. Another pipeline, passing in part through Lebanese territory, was closed down in 1976 because of the Lebanese civil war. One of the two pipelines was activated during 2000 but only to Syria. The Iraqis are trying to renew the supply to Lebanon but have not succeeded so far.

In the past, land transportation through Turkey to and from Iraq was relatively unimportant. But with the completion of the Kirkuk-Dörtyol pipeline in 1977 and especially following the outbreak of the Iran-Iraq War in the early 1980s, this land route assumed vital importance for Iraq.

The Kurdish problem, the allocation of Euphrates water, and, later, land transport have been closely intertwined with Turkey-Iraq oil relations. These issues have given Turkey some leverage against the huge advantage the Iraqis have had in terms of oil. By being able to offer Iraq something in three areas directly affecting Iraqi vital interests, Turkey has been able to obtain special terms in its oil supply, as related earlier.

Indicative of the intertwining of these issues—oil, water, and security—was the agreement made between Iraq and Turkey when Prime Minister Bülent Ecevit met with Iraqi vice president Saddam Hussein on 2 July 1974. It was agreed, as noted previously, that Iraq would provide Turkey with a loan for both oil purchases and the construction of its part of the pipeline, whereas Turkey undertook to increase the flow of water into Iraq and to coordinate with the Iraqis on matters concerning pipeline security, which meant in effect responding to Iraqi calls for help in protecting the flow of oil in the Iraqi part of the pipeline.

This triple deal increased mutual trust and cemented Ankara's relations with Baghdad. When the Süleyman Demirel government came into power in April 1975, it reaffirmed the agreement. Demirel also regarded Iraq as Turkey's oil supplier of choice in view of both geographic proximity and the pipeline, which was already under construction. Despite the numerous difficulties in Turkey's relations with Iraq in the 1970s following the oil price increases, the Kirkuk-Dörtyol pipeline remained beneficial to both parties, and the oil agreement signed in August 1973 has remained in effect.

In the late 1970s and into the 1980s relationships grew closer still with the barter agreement in August 1978, the increase in Turkish exports to Iraq, and the building of a second pipeline from Iraq to the Mediterranean through Turkish territory. At the same time, however, problems appeared in other areas of Turkey's relations with Iraq. Kurdish activities along their common border intensified, the water issue became more acute, and—most important—Iraq changed its attitude on Cyprus and began supporting the Greeks. The Iran-Iraq War added further problems, despite its overall positive effects on Turkish-Iraqi relations. For example, the Iraqis attacked several Turkish tankers in the Gulf, much to everyone's embarrassment.

Even when the tables were turned economically after the last energy crisis and Iraq began accumulating a huge financial debt to Turkey, the political aspects of the relationship remained basically the same. The Turkish press even claimed Turkey was exerting insufficient pressure on Iraq to repay its debt because it needed Iraqi cooperation on the Kurdish issue. But Turkey-Iraq relations met their most severe test when the Gulf War crisis began in August 1990.

Turkey and the Gulf War

When Iraq invaded Kuwait on 2 August 1990, Turkey's relations with Iraq—among other things—were thoroughly upset. Many Middle Eastern alliances were realigned as a result of this move by Saddam Hussein, including Turkey's hitherto close relationship with Iraq. It was obvious to the Western powers from the start that no significant move could be made against Iraq (initially it was thought that economic sanctions, up to a total trade embargo, would suffice) without active Turkish support. At the time, around 60 percent of Iraq's oil imports were flowing through Turkey, as well as a major part of its foreign trade in other commodities and goods.

Turkey barely hesitated before joining the sanctions. Even though such a move would clearly entail heavy economic costs for Turkey, political and strategic considerations outweighed economic ones. On 7 August Turkey's president, Turgut Özal, decided to close the Iraqi border to all trade, including oil.[3] Potential losses were numerous: the loss of passage fees on goods and oil, stoppage of all exports to Iraq, withdrawal of Turkish contracting firms from projects in Iraq, and losses because of the increase in global oil prices as a result of the crisis. Also,

Iraq immediately stopped all payments on its huge debt to Turkey. At the same time, Turkey had to seek alternative sources of oil to replace the more than 60 percent of total oil imports supplied by Iraq.

This is not to say that political considerations were simple or clear-cut. Since the mid-1960s, as we saw, Turkey's ties with Iraq had been a major element in Turkey's foreign relations, which contributed greatly to its economic recovery during the 1980s. Also, around 1.5 million ethnic Turks were living in Iraq—a majority of the population in the Kirkuk region—and Turkey could not ignore them. And the two countries had cooperated until then on the Kurdish problem, a source of grave concern for both. On the other hand, relieved of its erstwhile economic and political dependence on the Arab nations, Turkey under President Özal was making every effort to move closer to the West in general and to the European Union (EU) in particular. Besides, Turkey was uneasy over Iraq's massive military buildup following the termination of its war with Iran.

Turkey's swift response to the U.S. request to block Iraq's outlet to the Mediterranean and join an economic boycott was interpreted in the West as a major move in Turkey's return to the pro-Western policies it had adopted during the 1950s. At that time, under Prime Minister Adnan Menderes, Turkey had been regarded as the West's closest ally in the Middle East. Now, when Özal promised not only to join the embargo but also to observe it to the letter—even before Turkey was promised any compensation for the expected economic losses—there was much relief in the United States. It is doubtful that such an embargo could have been effective without wholehearted Turkish support.

It should be recalled that ten years earlier the United States had made a similar request to Turkey regarding Iran during the Iran hostage crisis. Then Ankara's reply was less than satisfactory to the Americans. But Turkey's economic fortunes had completely changed during the intervening decade. Now, faced with the difficult choice between suffering short-term economic losses and appearing to support Saddam Hussein against both the West and much of the Arab world, Turkey made the right decision as far as the anti-Saddam coalition was concerned.

Indeed, Turkey was strict in its implementation of the embargo. The oil pipelines leading from Iraq to the Mediterranean were closed, foodstuffs en route to Iraq were impounded, and ships carrying cargo destined for Iraq were not allowed to unload at Turkish ports. Turkey even offered Iran a $400 million credit to help that country cope with the hardships created by Turkey's participation in the embargo.

This was just the first stage of Turkey's involvement in the Gulf crisis. When it began to appear that the situation might deteriorate into a full-fledged war, the Turkish parliament approved by a large majority a bill granting the government full power to send troops abroad and to allow foreign forces to be stationed in Turkish territory. Thus Turkey joined the multinational force and offered it an option to open a second front against Iraq from Turkish territory. The bill also enabled the multinational force to use the North Atlantic Treaty Organization's (NATO) Incerlik base near Adana for operations against Iraq. Politically, Turkey repeatedly demanded that Iraq withdraw its forces from Kuwait; militarily, it concentrated forces—up to 112,000 troops—along its border with Iraq, forcing the Iraqis to divert troops from the main theater of war in the south. Yet Turkey announced that it would not directly attack Iraq unless it was attacked.

In early January 1991, a few days before the war broke out, Turkey had approached NATO headquarters in Brussels to request that NATO troops be deployed along its border with Iraq. Among other things, the Turks asked for 42 jet fighters to be sent there to supplement their own 150 planes. Three U.S. squadrons, with 48 F-15 and F-16 jets sent from Europe, were based in Incerlik when the war started in mid-January 1991.

Throughout the crisis President Özal made tremendous efforts to minimize the damage to Turkey's economy. The United States, Japan, Kuwait's government in exile, and Saudi Arabia promised compensatory assistance; the Saudis even offered oil deliveries under preferential terms, such as Turkey had sought in vain fifteen years earlier. Trade with Iran was expanded, and Turkey approached the USSR to find an outlet for its exports to replace Iraq. But even as economic damages were mounting, the chief consideration remained political. Ankara's policy was based on hopes of political support from the West once the Gulf crisis was over. The Turks wanted their request to join the EU to be considered favorably. They also sought a more evenhanded Western attitude toward their conflict with Greece over Cyprus. And there were other issues on which international opinion mattered a great deal to Turkey, such as regional water rights and the Kurdish problem.

Özal's wholehearted support for UN policies in the Gulf was also costly in terms of Turkish domestic politics. Those opposing his policies (mainly from social-democratic quarters) criticized both the president's assumption of final authority over foreign policy, bypassing both parliament and the government, and his willingness to involve Turkey directly in the war by becoming the staging area for a second front against Iraq. Özal also had problems within his own party and with both the military and the civil service. Foreign Minister Ali Bozer resigned when Özal returned from Washington in October 1990. In November the minister of defense resigned, and in December the chief of staff followed suit. The generals were concerned that Turkey would be left alone to face Iraq once the coalition withdrew its forces. Prime Minister Yıldırım Akbulut vacated his post in early 1991. The clergy was also unhappy over the prospect of war with another Muslim nation. Much of the media reflected such criticisms. Thus Özal was weakened domestically—even some of his supporters regarded his moves as a dangerous gamble—while increasing his own and Turkey's prestige abroad and securing Western commitments for Turkey.

When war broke out on 16 January 1991, however, Turkey became fully involved. The next day parliament approved the government's request not only to allow foreign troops stationed on Turkish soil to fight against Iraq, using the Incerlik air base, among others, but also to send Turkish troops to the front. Turkey asked the United States to send forty-eight more jet fighters, in addition to the forty-eight planes already there. The civilian population in the southeast part of the country was preparing to absorb bombardments by Iraqi Scud missiles, possibly armed with unconventional warheads.

On 17 January, at an emergency meeting in Brussels, the NATO governments promised to defend Turkey should the need arise. That same day Iraq's foreign minister sent a letter to his Turkish counterpart, describing Ankara's moves as aggressive and threatening dire consequences. Turkey sent its foreign minister to a series of Arab capitals to explain that Turkey had never had territorial ambitions in Iraq, would not use water supply as a means of pressure, and would desist from interfering in Iraqi internal affairs.

On the third day of the war Incerlik was used to launch dozens of U.S. aircraft to attack targets in western Iraq. U.S. command and control aircraft also took off from Turkish territory. In fact, Turkey became much more actively involved in the war than many other NATO members. After the United States and the United Kingdom, Turkey can be regarded as the most significant member of the anti-Saddam alliance during the hostilities.

The Aftermath of the Gulf War

The Gulf crisis caused Turkey to depart from two principles that had guided its foreign policy for decades: neutrality in intra-Arab conflicts and good relations with Iraq. Chief among the causes that forced Turkey to rethink its policies in August 1990 was the simple fact that it could not acquiesce with Iraq's conquest of Kuwait. For Turkey, this was not only a gross violation of international norms but also a possible precedent, with worrisome long-term implications. Also, Turkey was looking for a new role in NATO: the Cold War was almost over, and Turkey was about to lose its importance as the linchpin of NATO's northern tier. With this decline in importance, decreased economic and military assistance was sure to follow. The Gulf crisis gave Turkey an opportunity to demonstrate its strategic significance to the West for the future as well.

That aim was fully achieved. Turkey's prestige in Washington and other Western capitals reached new heights. Western willingness to help renovate Turkey's military, which had been almost nonexistent since the Cyprus crisis broke out in the 1970s, was revived.

On 7 April 1991 U.S. secretary of state James Baker came to Ankara for a twenty-four-hour visit, the first high-level meeting since the end of the war. At this time and subsequently, Turkey was promised an almost undreamed-of package to be delivered over the next few years: 600 M-60 tanks, 400 Leopard tanks, 700 armored personnel carriers, 40 F-4 Phantom jets, Cobra assault helicopters, and Roland surface-to-air missiles. President George Bush also agreed to consider allowing Turkey to assemble F-16 jet fighters, with possible Saudi participation. He also ordered that Turkish textile import quotas to the United States would be doubled as a show of gratitude. In all, Turkey was to receive $8 billion in military supplies, mainly from the United States and Germany.

It is estimated that Turkey's economic losses during the first year of the crisis amounted to $9 billion, whereas direct economic compensation reached only $2.2 billion, received as grants, loans, and cheap oil from Saudi Arabia, Kuwait, the United Arab Republic, Japan, France, and the European Community.[4] Turkey did win a favorable position for its contracting companies in reconstruction efforts launched by Kuwait, and its improved political relations with both Iran and Syria were expected to yield some economic advantages as well. The economic balance, however, was still in the red. The real compensation was a strengthened Turkish military, as well as an enhanced international position.

On 12 June 1991 Iraq's deputy prime minister, Tarik 'Aziz, went to Ankara. It was the first time an Iraqi official had visited one of the nations that had participated in the coalition war against his country. Three months after the end of hostilities, 'Aziz expressed willingness to renormalize Iraq's relations with Turkey on the basis of good neighborly relations and noninterference in internal affairs. The Turks replied that they were willing to reestablish relations with Iraq but only on the basis of UN resolutions. This meant, in effect, that economic relations could not be resumed, as Iraq was still under economic sanctions.

Internationally, then, Özal's gamble was vindicated. But domestically his Motherland Party paid a heavy price: it lost the 1991 election. Özal remained president, but his rival, Süleyman Demirel, became Turkey's new prime minister. And it was Demirel—this time as state president—who eight years later told the new Iraqi ambassador in Ankara, "Turkey has no problems with Iraq, but Iraq has many problems with the world. First you have to eliminate these problems. Until that time, there is not much to do."[5]

Generally speaking, Turkey had followed the UN position on Iraq, supporting the international effort to defuse Iraq's mass destruction weapons and demanding that positive steps be taken toward greater democracy inside Iraq. Still, the main items on Turkey's agenda were the curbing of PKK activities and the need for compensation for the economic losses it suffered because of the international sanctions against Iraq.

The struggle against Kurdish insurrection, which made Turkey carve a security zone for itself in Iraqi territory, is discussed later. On this, Turkey needed Iraqi cooperation, which was not readily forthcoming. At the same time, a fierce debate was taking place domestically about Turkey's role in the ongoing effort by the allies to monitor military activities in northern Iraq. Specifically, every six months the Turkish parliament was asked to renew the mandate given to the Northern Watch Force (composed of U.S., British, and Turkish elements) to use the Incerlik air base for its operations. Things came to a head in December 1998, when the allies decided to launch Operation Desert Fox against Iraq because of its failure to abide by UN resolutions. This time Turkey flatly refused to allow its territory to be used for operations against Iraq.

Economically, the ongoing embargo against Iraq throughout the

1990s caused Turkey substantial financial damage, mainly because of the suspension of operations of the oil pipelines through its territory and the loss of Iraqi oil supplies.[6] Saudi Arabia took over as Turkey's chief oil supplier in the 1990s, followed by Iran, the United Arab Emirates, and Libya, but that oil had to be hauled over longer distances and was therefore more expensive. Understandably, then, Ankara's official position held that some action had to be taken by the West to allow an end to the embargo.

Iranian Fluctuations

Between 1969 and 1972, Ankara made strenuous efforts to persuade Iran of the feasibility of a pipeline from Iran to the Mediterranean Sea through Turkish territory. The plan made great economic sense for Turkey, which was importing increasing amounts of oil, and Turkey also liked the idea of earning passage fees. The plan was also sensible for Iran, because it could reduce transportation costs to European ports (the Suez Canal was blocked at the time as a result of the Six Day War). Politically, it made even greater sense: Turkey needed Iran, the only other major non-Arab nation in the Muslim Middle East, as a close ally. Thus at the same time Turkey was making great efforts to persuade Iran not to withdraw from the Baghdad Pact alliance[7] and to take a more active role in the Organization for Regional Cooperation for Development (RCD).[8] Between them, these organs were supposed to link Turkey and Iran to each other and to the West. An Iran-Turkey pipeline could have given much substance to relations between the two countries.

The idea of a pipeline from Iran to a Turkish Mediterranean port dates back to 1956, when the Suez Canal was closed for a short time after the Suez Campaign. It was revived when the canal was closed for a much longer period after 1967. At the time, maritime transportation costs were increasing all over the world, and Turkey was beginning to appreciate that its energy needs in the near future were going to be much greater than before because of its rapid industrialization. In 1969 Turkey and Iran decided to establish a joint Oil Transport Authority, charged with planning a pipeline from the Ahwaz oil fields in Iran to the Turkish port of İskenderun.[9] Negotiations went on for several tedious years until they collapsed in early 1973.

Several obstacles accounted for this failure. Iran refused to commit to a minimum amount of flow, thus making viability calculations virtually impossible. It also demanded to receive the Mediterranean (rather than the Gulf) posted price for oil at the pipeline's outlet in İskenderun; this meant Turkey would have to forgo some of its passage fees to keep the price of the oil competitive. Also, maritime transportation costs began to go down in the early 1970s. Additionally, Iran was showing great interest in an alternative proposal—the Suez-Mediterranean (SUMED) pipeline, intended to be built along the blocked Suez Canal between Suez City and Port Said.

When negotiations collapsed, the shah said economic considerations had caused Iran to prefer shipping its oil in tankers rather than through a pipeline. But Turkey felt the real motives were political in nature. The Turks, who had already suspected the shah of avoiding cooperation with them because he regarded Turkey as a rival for regional hegemony, were left with a bitter taste.

With the 1973 oil crisis and the unequivocal Saudi refusal in early 1974 to sell Turkey any oil, Ankara had little choice but to resume its overtures toward Iran. Despite previous disappointments, the Turks believed Iran would eventually come around and sell them oil and, more important, agree to a proposed natural gas pipeline through Turkey to Europe that would enable Turkey to draw some gas for itself. Throughout 1974, bilateral diplomatic relations between Ankara and Teheran consisted of negotiations over the gas pipeline, to the exclusion of almost everything else. The Iranians were undecided, because they also had the option of building a pipeline through Soviet territory, but they preferred to lead the Turks on, claiming repeatedly that it was their full intention to have their pipeline reach one of Turkey's Mediterranean ports.[10] In May 1974 the Turkish Foreign Office confirmed that Iran had signed a gas pipeline agreement with the USSR but was also considering another project that would deliver liquefied gas to the Balkans through Turkish territory. Energy Minister Chaıt Kayra said the pipeline would benefit Turkey in several ways: through passage fees, the use of gas as fuel, the development of petrochemical and fertilizer industries, and increased power production in the southeastern part of the country.[11] He added that upon completion of the project, İskenderun would become a major industrial center.

But in July 1974 a note of skepticism became noticeable in the Turkish press. A suspicion arose that the Iranians were using the option of building their pipeline on Turkish territory as a bargaining chip in negotiations with the Soviet Union over a similar project.[12] The Iranians would not give a definite answer when asked about the pipeline's future, and Ankara was becoming increasingly concerned. When Iranian foreign minister 'Abbas Khal'atbari visited Ankara on 18 January 1975, he explained to President Fahri Korutürk and Foreign Minister Melih Esenbel that his country's considerations in this matter were purely economic in nature. The Turks realized that they had to make an urgent plea to Teheran—"at a high level and of a political nature"—if they wanted to save the pipeline. But even that did not help. By late January the feeling in Ankara was that the gas pipeline project was "a total loss."[13] Further appeals by the foreign minister in March and the president in June 1975 only confirmed that belief.

With the failure of negotiations over an Iranian gas pipeline—an almost exact repeat of the failure of the oil pipeline negotiations three years earlier—Turkish frustrations exploded. Once again Iran had led Turkey down the garden path, procrastinating and making false promises and eventually explaining away its decision to abandon the project with economic cock-and-bull stories. This only added to the resentment already felt over Iran's refusal to give Turkey special treatment in oil sales. Early in 1975 the Turkish press vented its accumulated outrage. Not by coincidence, this happened right after Turkey had signed its agreement with Libya and was feeling more secure about its oil supplies. Stories appeared chastising Teheran for undertaking extravagant defense expenditures, for attempting to isolate Iraq in the Muslim world, for providing oil to Israel, for failing to develop trade with Turkey, and even for torturing prisoners in its jails. Editorials described Iran's fears over Turkey's strengthened regional status and mentioned the possibility that Iran and Turkey would part ways once and for all if the gas pipeline project were dropped for good.[14]

Some newspapers called on the government to retaliate. One commentator, *Cumhuriyet*'s Ümit Gürtuna, stated that Turkey had received from Iran "not the slightest measure of friendship" and went so far as to warn of the possibility that Iran would "turn its guns against Turkey, once it has settled its account with Iraq." With the rapprochement between Turkey and Libya, even the left-wing *Cumhuriyet* could find ideological justification for Turkey to turn its back on Iran. Facing the "reactionary triangle" consisting of Saudi Arabia, Iran, and Egypt under U.S. auspices, the newspaper wrote, a new "reformist triangle" has emerged consisting of Turkey, Libya, and Iraq. Should Turkey purchase its oil from Libya rather than Saudi Arabia or Iran, it would be helping a "modern" nation rather than a backward one.[15]

Ankara's politicians shared this anti-Iranian outrage but were more reluctant to express it publicly. Haluk Bayülken, the foreign minister when the gas pipeline negotiations started, later said: "The shah's refusal angered us because it was so opportunistic. He was tied to the West, but he preferred to have his pipeline go through the Soviet Union rather than Turkey, a nation that has sacrificed so much for the sake of regional security. The shah's motives were in the main strategic and political. His fear of strengthening Turkey once again played a major role."[16]

When Süleyman Demirel's government came into power in March 1975, Turkey's relations with Iran were reexamined. Concerned about Turkey's deepening relations with Libya, Demirel decided to give Iran "a second chance," despite the series of blows Iran had dealt during previous years. Endeavoring to end Turkey's exclusive reliance on Arab oil, Demirel decided to resume economic relations with Iran.

In June 1975 Turkey's President Korutürk visited Teheran. During that visit the two nations entered into an economic and technical cooperation agreement. Iran agreed to give Turkey a $1 billion "soft" credit for twenty years, an offer greeted with much enthusiasm in Ankara. Foreign Office officials described the agreement as "the best ever made with Iran."[17] The press spoke of "a new era" in Turkey's foreign relations,[18] and Prime Minister Demirel told his ministers on 10 June 1975 that the development of Turkey's relations with Iran was "one of the highlights" of his government's program.

Despite all the excitement, the agreement never lived up to Turkish hopes. The main obstacle, now as ever, was the shah's refusal to sell Turkey oil under preferential terms, a move regarded by Ankara as an indication of his true intentions toward Turkey.

Yet another attempt to cement Turkey's economic and political ties with Iran failed the next year. At the RCD summit in İzmir in April 1976, President Korutürk suggested the establishment of a common market for Iran, Turkey, and Pakistan based on their geographic proximity, shared view of the world, and common economic interests. The shah, as usual, was less than enthusiastic, and the suggestion was dropped.

U.S. Hostages in Teheran

Turkey's relations with Iran during the years 1974–1977 can only be described as chilly, but in July 1978, as related earlier, the two nations entered a large-scale barter agreement on oil imports. Relations began to improve, and when Ruhollah Khomeini seized power in Teheran a few months later, the situation changed completely. Then, on 5 November 1979, the protracted drama of the U.S. hostages in Teheran began. The U.S. appeal to Turkey to join in the economic embargo it had imposed on Iran put the improved relations between Ankara and Teheran to their first major test.

As soon as the crisis began, Ankara realized it had to walk very carefully on the tightrope between "a neighbor and a friend" (Iran, at least since the July 1978 agreement) and "an ally" (the United States), as Foreign Minister Hayrettin Erkman put it to parliament.[19] Turkey needed Iran to overcome the oil crisis, yet it also needed the United States to revamp its defense system, much weakened following the deterioration of U.S.-Turkish relations in the second half of the 1970s. Then again, in matters of national security, Teheran had played an important role in Turkey's struggle against Kurdish irredentism.

Early on, the Iranians found it necessary to explain that Turkey's participation in a U.S. economic embargo would hurt its oil supplies and could even bring about a severance of diplomatic relations. Ankara could not afford such censure and hastened to reassure Teheran. In a joint session of both houses of parliament on 23 January 1980, Foreign Minister Erkman announced that Turkey did not believe economic sanctions would contribute to the resolution of the crisis and thought they might even exacerbate existing misunderstandings. Erkman stressed that Turkey had no intention of imposing sanctions on the Iranian people. A few days later, the foreign minister delivered a formal message to that effect to Iranian representatives.

Happy with Turkey's initial reaction, Teheran nevertheless feared that a combination of U.S. and Iraqi pressure could make Turkey change its mind. To counter this possibility, it decided to give the Turks economic reasons to stick to their guns. On 28 January 1980 Iran's president-elect, Abu al-Hassan Bani Sadr, told *Milliyet* that Iran was preparing an economic assistance package for Turkey, intended "to free it from the pressures exerted by foreign banks."[20] Shortly afterward Iran announced its decision to set up joint agricultural and industrial projects and its willingness to fulfill all of Turkey's energy needs.[21] Coming as they did in the midst of Turkey's worst economic crisis in its history, these announcements were a boost for Ankara.

In early April 1980 U.S. secretary of state Cyrus Vance asked several Western ambassadors in Washington, D.C.—including Turkey's—to support the embargo his country intended to impose on Iran.[22] A few days later the U.S. ambassador in Ankara served notice of the forthcoming embargo in a letter to the Foreign Office.

Turkey's reaction was swift. Acting foreign minister Ekrem Ceyhun called the notion of economic sanctions against Iran "inconceivable" because of what he called "the historical ties" between the two nations.[23] Other officials spoke of the "special relationship" between Iran and Turkey as ruling out any possibility that Turkey would join.[24] Later, the Foreign Office made public its official reaction to the U.S. request. It said the nature of the "special relationship" between Turkey and Iran had been explained to the Americans, and the parties had reached an understanding regarding the "special aspects" of relations between Ankara and Teheran. The posture taken by Demirel's government enjoyed the support of opposition leader Bülent Ecevit. In a statement on behalf of his party, Ecevit rejected any attempt by the United States to dictate to Turkey the way it should deal with a neighboring nation and expressed his fear that Turkish participation in the embargo would push Iran into the Soviet camp. The Turks also realized that a Western embargo would open new markets for Turkish exports in Iran. At the same time, Turkey would also benefit from passage fees on goods delivered to Iran by land if Iran's seaports were blockaded.[25]

Thus Turkey became the first NATO member to announce officially that it would not join the economic embargo against Iran. Still, the Turks' attitude toward the political aspects of the crisis was completely different. The general public, the media, and the political system were outraged by the taking of U.S. hostages. Although such sentiments could not be expressed at the bilateral level because of the complexity of Turkish-Iranian relations, on a broader political level the Turks realized the need to maintain solidarity with the United States. As a practical expression of that attitude, Foreign Minister Erkman announced on 24 April 1980 that Turkey would not allow the transfer through its territory of goods to Iran originating in countries that did join the embargo.

The Iranians kept encouraging Turkey's objection to the sanctions. At the same time, they also tried to persuade Turkey to reduce its political support for the United States in the crisis. Emphasizing their common religious and cultural backgrounds, as well as geographic proximity, Iran endeavored to convince Turkey that its foreign policy should be more "orientally oriented." Iranian foreign minister Sada Ghotbazadeh (in an interview with *Milliyet* on 15 April 1980, four days before Turkey made its official statement) called on the Turkish government not to join the U.S. embargo. Such a posture, he said, would enhance Turkey's prestige, emphasizing its independence from the great powers. He added that Iran would regard such a refusal as a "show of real friendship" by Turkey at a time when external pressures were being exerted to isolate Iran from the world. Ghotbazadeh used the opportunity to reaffirm Iran's willingness to help Turkey economically and to cooperate with it in international Muslim organs.

The Iranian chargé d'affaires in Ankara, Ahmad Mossavi Zade, also described the Turkish position on the embargo as a test of the friendship between the two nations, as well as of Turkey's ability to protect its national interests against Western pressure. He saw fit to remind the Turks that whereas during the shah's regime Turkey had received from Iran "not a single drop of oil," Khomeini's Iran had made strenuous efforts to aid Turkey, providing it with 2.4 million tons of oil a year.[26] When Iranian president Bani Sadr was interviewed by *Cumhuriyet* on 10 May 1980, he emphasized the two nations' "common destiny" as a good enough reason for them to "collaborate against Western hegemony."

The Iranians kept up this extraordinary display of friendship and willingness to help during May 1980. They repeatedly promised to meet any Turkish request for oil and declared their intentions to expand imports from Turkey and increase the number of joint ventures. At least as far as oil was concerned, this was not idle talk. When Turkey announced its position on the U.S. embargo in January 1980, the Iranians began shipping crude oil and distillates without demanding cash payments. Turkey's oil debt to Iran, which stood at $77 million in February 1980, rose to $277 million by July of the same year. At that time Turkey had nearly exhausted the $300 million credit line opened for it by Iran.

Turkey was careful to adhere to the policy guidelines it had set for itself when the crisis began. It refrained from engaging in any anti-Iranian act on the bilateral level but supported as a matter of principle the U.S. position on the hostage affair. In effect, Turkey was walking a very narrow path between the constraints imposed by its relationship with Iran on the one hand and its wish not to alienate the United States on the other. Having managed to persuade Washington of the delicacy of its relations with Iran, Turkey had been able to improve considerably its economic relations with Teheran without prejudicing its political relations with the Americans. Needless to say, Iran's financial generosity, both actual and promised, had a great deal to do with the posture adopted by Turkey's leaders in the crisis.

Relations with Iran
During and After Its War with Iraq

The 1980s were years of war between Iran and Iraq, and again Turkey had to adjust its policies toward its two neighbors according to the changing fortunes of war. From the very beginning Ankara called for a swift end to the war, but it refused to assume any role as a mediator except for its participation in a goodwill mission, sent by the Islamic Conference to offer its good services to the belligerents. Turkey adopted a position of strict neutrality, becoming involved only in such humanitarian ventures as sponsoring an exchange of prisoners of war (1983) and an exchange of diplomats (1984).[27]

The economic impact of the Iran-Iraq War on Turkey has been discussed extensively (see Chapter 5). As a result of the expanded bilateral trade and the blockade on Iranian seaports, a large part of Iran's overall foreign trade (except oil) went through Turkey during the 1980s. In February 1983 the two nations decided to keep their chief border crossings open around the clock rather than eight hours a day as before. The number of vehicles moving through these points soon rose from 600 to 1,200 a day. Several Turkish ports, particularly Trabzon and İskenderun, benefited greatly from the added business.

Yet the war was not totally a blessing for Turkey. One of the main problems it created was a deterioration of the safety of maritime passage in the Gulf, as well as of the security of land transportation of oil through Turkish territory. Ankara was worried about Turkish tankers plying the Gulf routes and noted with concern the Iranians' public threat to hit the Kirkuk-Dörtyol pipeline, which they declared to be a legitimate strategic target. That did not happen, but during the war ten Turkish tankers were hit—eight by Iraq and two by Iran.[28]

Yet another problem had to do with Turkey's diplomatic relations with Iran. Many Iranians had left their homeland because of their opposition to the Khomeini regime and had found their way to Turkey, where they were settled in camps in the eastern part of the country. Some were shah loyalists who wanted to go to the United States or Europe and were allowed to do so. Others, however, attempted to launch subversive activities from Turkish territory across the border. Turkey forbade Iranian opposition members to organize in its territories under threat of jail or deportation.[29] By 1985 the number of Iranian refugees in Turkey had reached nearly 1 million, and they had become a real economic burden.[30]

At the same time, throughout the 1980s there was an intensification of Kurdish insurrection against Turkey from Iran's territory. Teheran had an obvious interest in a Kurdish rebellion against Iraq as an ongoing military threat. The Kurds, however, did not care whom among their oppressors (as they saw them) they hit. Furthermore, any Kurdish success in one area was sure to encourage activities in other places. In view of all these problems, Turkey and Iran agreed to announce simultaneously that neither country would allow its territory to be used as a basis for insurgency against the other.[31]

Islamic activity, however, was not regarded as insurgent, at least not by Khomeini's Iran—which, as noted earlier, had become deeply involved in such activities within Turkey. This was a supreme policy goal for Iran, and it did not hesitate—despite its growing dependence on Turkey in the economic sphere—to express its displeasure with Turkish secularism. Thus when Prime Minister Hossein Mussavi went to Ankara in June 1987, he made a point of not visiting the Atatürk mausoleum, the shrine of Turkish secularity; he went even further, publicly criticizing the great leader's secular philosophy. The Turks refrained from comment.[32]

The Struggle for Central Asia

In the early 1990s a new and very significant dimension was added to relations between Turkey and Iran with the emergence of six new Muslim former Soviet republics in their vicinity. Both nations began efforts to shape the future course to be taken by Kazakhstan, Uzbekistan, Tajikistan, Kyrgyzstan, Turkmenistan, and Azerbaijan. Turkey has offered guidance in the areas of democracy, secularity, modernization, and Western affiliation. Iran has offered the Islamic alternative, appealing to those peoples' religious sentiments. Turkey has enjoyed an advantage in this race, both because of cultural and linguistic affiliations and because it can act as a bridge between these nations and the West.[33]

Turkey conducted its relations with these republics under the motto "help us help you," as President Özal put it.[34] The official position was this: if the republics chose Turkey's way, Turkey would be willing to help them. The Turkish model defeated the Iranian one in five of the six central Asian republics (only Tajikistan is closer to Iran than to Turkey). It soon became apparent that the republics were more interested in rapid economic development, particularly for their oil and gas sectors, than in Islamic affinities. The Caspian Sea is known to have huge reserves of oil and gas, and the republics surrounding it—Azerbaijan, Kazakhstan, and Turkmenistan—needed massive financial investments to utilize those reserves. For them, the Turkish model was the key to Western capital.

A collateral interest for these republics, most of which are land-locked, has been a route for their oil and gas to reach Western consumers. In 1992 Azerbaijan and Kazakhstan announced that they preferred to ship their oil through Iran. A high-level U.S. intervention, however, in which President Bill Clinton became personally involved, persuaded the leaders of the two nations to change their minds in favor of Turkey.

With the election of Vladimir Putin as president of Russia in early 2000, the Russian attitude toward the former constituents of the USSR became more assertive. During the 1990s competition for political and economic influence in the six republics was mainly between Turkey and Iran, but Russia now offers them a third alternative.

Although Turkey easily defeated the Iranian challenge in terms of general influence and provision of a role model, it has not gained much of an economic and political foothold in these central Asian nations. Their major contracts and energy concessions have gone largely to U.S., European, and Japanese companies. Still, these nations remain a top priority for Turkish foreign policy, and it maintains efforts to keep up its competition with Russia.[35]

Iran and Turkish Secularity

As has been noted elsewhere in this book, postrevolutionary Iran took it upon itself to bolster the position of Islam in Turkey. Until the late 1990s that effort consisted mainly of financial support to religious organizations, as well as a few symbolic gestures—annoying but inconsequential for official Turkey. There was little attempt to interfere directly in Turkish politics.

That changed abruptly in February 1997. In the town of Sincan outside Ankara, the religious Welfare Party organized a "Jerusalem Night" in protest of perceived Palestinian concessions in the Middle East peace process. The affair took on a distinctly antisecular character, with slogans such as "a true Muslim fights to the end" and "the foundations of

Shari'a will be laid here."[36] The guest of honor at the event was the Iranian ambassador to Ankara, Muhammad Reza Baghri.

The military, ever watchful for antisecular demonstrations, reacted swiftly. Twenty tanks rolled into Sincan, carrying a clear message: no threat to the secular order would be tolerated. The Iranian ambassador was asked to leave Ankara at once. All of this evoked memories of an incident seventeen years earlier, when an Iranian ambassador participated in the demonstration that preceded the September 1980 coup d'état.

Domestically, this development started a chain of events that led to the removal of Prime Minister Necmettin Erbakan from power, primarily because of his pro-Islamic proclivities, as discussed later. One of the demands the National Security Council made to Erbakan prior to his removal related specifically to Iran: "The activities, attitude and conduct of the Iranian Islamic Republic against our country's regime must be obstructed. . . . A package of precautionary measures to prevent deliberate and injurious activities must both be prepared and applied."[37]

The Sincan affair brought Turkish-Iranian relations to a near standstill. The economic component of those relations, however, was too important for both sides to forgo. Also, Turkey could not afford to ignore the constant threat from the PKK, which could have become much more serious had Iran chosen to give it full support. Thus in September 1997, two months after Erbakan's resignation, the foreign ministers of the two nations met in an effort to renormalize relations. A major obstacle was Turkey's relationship with Israel, which the Iranians bitterly resented. The Turks, however, brushed off protests by the Iranians (which were numerous), claiming their alliance with Israel was not directed against anyone.

In June 1998 security relations between Turkey and Iran were resumed with a meeting of the joint Security Commission, the first of its kind in two years. Just two months later Iran offered its services as a mediator in the Turkish-Syrian dispute over Syria's support of the PKK, discussed later.

The period of renormalization, however, was short-lived. On 15 May 1999 the Turkish Foreign Ministry found it necessary to demand again—publicly—that Iran stop interfering in Turkey's internal affairs. Not only was Iran providing financial aid to Turkey's religious party (now called Fazilet Partisi, the Virtue Party), it was also providing shelter to PKK guerrillas and was even training members of a Turkish insurgent group with an Islamic orientation, the Hizballah.[38]

Turkish reaction was harsh. Prime Minister Ecevit accused Iran of having taken over the role previously played by Syria in aiding and abetting the PKK and other Turkish insurgents, and he called on the Iranians to desist immediately.[39] His warning was ignored. In mid-2000 ten Hizballah members were caught and charged with a series of assassinations of prominent Turkish liberals. There was no doubt that Iran was supporting them clandestinely. During the summer of 2000 more and more information came to the fore exposing Iranian subversive activities. At the time of writing, relations between the two nations had reached rock bottom.

The Special Relationship with Libya

In the early 1970s disappointment over both the shah's evasiveness and Saudi Arabia's cold shoulder led Turkey to develop relations with Libya, whose political and military support during the Cyprus War was the turning point in that decision. When Turkey invaded Cyprus in July 1974, it was denounced by most nations, and the UN organization condemned the intervention. Most Muslim nations took a guarded position, but Libya and Iraq were conspicuous in their support of Turkey. Besides logistic assistance provided by both countries—Libya with military supplies and fuel, Iraq with crude oil and jet fuel—they also extended diplomatic support. General Emin Alpkaya, commander of the Turkish air force, said Libya's help was "the stuff of legends."[40] He said Libya had offered Turkey the use of its Mirage aircraft and even its only F-5 squadron at a time when Belgium, West Germany, and the United States were withholding military supplies scheduled to be delivered according to previous agreements. Thus the Libyan offer was a boost to Turkey's morale as well as its military strength. In early September 1974 the Libyans told Finance Minister Deniz Baykal, in Libya to explain his country's policies, that Libya would assist Turkey in its diplomacy regarding Cyprus, including the use of Libya's good offices with other Arab nations and at the United Nations.[41]

Libya's favorable attitude during the Cyprus crisis had far-reaching effects. Turkey at the time was ruled by a socialist government whose leaders had fewer qualms than their predecessors about moving closer to pro-Soviet nations and had already expressed support for the policies adopted by the Organization of Petroleum Exporting Countries (OPEC). Those leaders, including Prime Minister Ecevit, his Deputy Erbakan, and Finance Minister Baykal, as noted earlier, believed the oil producers had long been exploited by the West and were entitled to retaliate. They also recognized that the increased power the producers enjoyed was a fact to be reckoned with, and during 1974 they tried to forge "special relationships" between Turkey and the producers, seeking to minimize the impact on Turkey of the greatly increased oil prices. These views, shared by most members of Ecevit's government, were popular with both Libya and Iraq but not with Saudi Arabia and Iran, which at the time did not intend to improve their relations with Turkey.

The success of this "moralist approach" toward the oil producers was quite pronounced, as related earlier, in the case of Libya. The Libyans found common ground with the leaders of the Cumhuriyet Halk Partisi (CHP, Republican People's Party), whose leaders missed few opportunities to highlight the similarities between their party's objectives and the declared principles of Libya's ruling party.[42] President Muammar Qaddafi, for his part, hoped to bring Turkey closer to the Arab world and to use it as a channel for furthering Libyan objectives in the West. All of this took place against a background of deteriorating relations between Turkey and the United States following the Cyprus invasion, which made Turkey even more dependent on Libyan aid and raised hopes in Libya for a real breach between Ankara and Washington.

The highlight of the mutual courting came in January 1975, when Libyan prime minister 'Abd-al-Salaam Jalud visited Ankara. Before Jalud left Tripoli his oil minister declared, "Libyan oil wells are put at the beck and call of the heroic Turkish people."[43] Upon his arrival in Ankara Jalud said both nations would work hand in hand from then on: "We are willing to make every effort necessary to mend the fences between the two nations."[44] A few days later, speaking of his country's assistance to Turkey during the Cyprus War, Jalud said, "We were willing to fight shoulder-to-shoulder with Turkey and share what we have with the Turkish people."[45]

During that visit Turkey and Libya signed four agreements for cooperation in economic, commercial, scientific, and cultural affairs and another on Turkish workers in Libya.[46] At the signing ceremony Turkish prime minister Sadi Irmak said, "Turkey and Libya have decided to unite their economies and their commercial and cultural resources for the benefit of both peoples."[47] Irmak added that in foreign policy there was also no conflict between the two; "on the contrary, there are parallel lines." Libya's prime minister added a few days later that "the peoples of Turkey and Libya have become one nation under Islam, and together they will write glorious and dignified pages of history."[48] Turkey's prime minister would not be outdone, saying: "Turkey has won a new friend and brother. We have made a decision to cooperate with the Libyan people all the way through. We shall put at Libya's disposal our trained manpower, and Libya will put at our disposal its rich resources."[49]

The Turkish press went overboard with praise for Libya when the agreements were announced. *Günaydin* wrote (6 January 1975) that Libya's revolutionary government could be regarded as "Islam's new pioneers." *Pulse* wrote in an editorial (same date): "Our NATO ally, the United States, refuses to sell us arms for which there are signed contracts. Our CENTO ally, Iran, would not sell us oil for the international price. It seems that the time of these alliances has passed. Turkish foreign policy must change, adjusting itself to the realities created by the ratification of the oil pipeline project with Iraq and the new agreements with Libya."

Turkey's chief motive for entering these agreements was economic. It needed to purchase oil under "soft" terms, and it needed an outlet for its surplus workforce—unemployable in the overcrowded Turkish market and now unable to find jobs in western Europe, where economic growth had slowed because of the energy crisis. Libya was willing and able to meet most of Turkey's requests in the economic sphere. The Libyans had a broader agenda in the political and military spheres and hoped to influence Turkey sufficiently to abandon its pro-Western foreign policy and cut off relations with Israel and Egypt. Tripoli hoped to make Turkey a member of the "refusal front"—those Muslim nations that objected to *any* dealings with Israel, let alone to peace with that country. Also, the Libyans hoped Turkey would become more involved in Muslim international affairs and in the political struggle for a new international economic order (NIEO). Thus the preferential supply of Libyan oil to Turkey was intended from the start to make Turkey dependent on Libya so the latter's objectives would be more readily attained.

The Libyans never hid or disguised their design but repeatedly stated their intent in public. President Qaddafi told Turkish journalists who accompanied Turkish deputy prime minister Erbakan on his visit to Libya in March 1975, "We are ready to fight alongside Turkey, shoulder to shoulder, in order to save our lands from Western enemies."[50] On another occasion he told Foreign Minister İhsan Sabri Çağlayangil: "We are happy with your strong reaction against the United States. It was joyously received by the Arab world. The Arab nations now respect Turkey all the more."[51] Militarily, Libya expected to use Turkish technological know-how to develop its military industries and acquire military hardware; it also wanted the Turkish army to train pilots and other personnel. There were economic expectations as well. Turkish skilled labor was intended to relieve Libya of its embarrassing dependence on Egyptian doctors, teachers, and engineers. In all, Turkey's expectations in entering its agreements with Libya were for the short term—it just wanted to put the present economic crisis behind it—whereas Libyan expectations were rather long term and were much more political in nature.[52]

The series of agreements entered into with Libya increased Turkey's confidence. Its policymakers believed economic reliance on Iraq (following the ratification of the oil pipeline project the same month) and Libya would secure a regular supply of crude oil and yield more economic benefits besides. Moreover, both Iraq and Libya agreed to sell their oil under preferential terms, which Saudi Arabia and Iran had persistently refused to do.

The situation barely changed when Süleyman Demirel became prime minister in April 1975. The Demirel government inherited accomplished large-scale contractual agreements with Iraq and Libya, and Demirel, like it or not, well knew how vitally important those agreements were for the Turkish economy. Politically, the agreement with Libya was more of a problem for Demirel than the Iraqi one. The leader of the Adalet Partisi (Justice Party) was not prepared to abandon Turkey's traditional pro-Western policy, as required in effect by the former agreement. He feared the Libyan agreement had thrown a monkey wrench into Turkey's already complicated relationship with the West. Still, in view of the new agenda the energy crisis imposed on Turkish foreign policy, he could not think of downscaling relations with Libya as long as a suitable replacement could not be found elsewhere among the Muslim oil-producing countries.

And indeed, the Libyans demanded full price for their willingness to rescue Turkey from its economic crisis. They exerted pressure far and wide, endeavoring to influence Turkey's relations with the United States, NATO, Israel, and Egypt for the worse and with the Palestine Liberation Organization (PLO) and other national liberation movements of the same ilk for the better. During the 1970s and 1980s Libyan leaders made numerous statements to that effect. A particularly patronizing example was Qaddafi's pronouncement to a group of Turkish journalists who went to Tripoli in July 1978, shortly after Turkey had allowed the PLO to open an office in Ankara: "We are not content with just the opening of a Palestinian office in Turkey. . . . Turkey should withdraw its NATO membership in order to assume its place in the Arab world and the third world. . . . Turkey and its people must not live by leave of the Americans. If I were a Turkish leader, I would have changed its foreign policy in this direction."[53]

During the 1980s Turkey's relations with Libya changed in line with the changing economic circumstances (as described in Chapter 5). On the one hand, the scope of projects undertaken by Turkish companies in Libya reached $9.4 billion in the late 1980s, and 60,000 to 80,000 Turks found employment in that country. On the other hand, the Libyans often failed to meet their commitments to the Turkish government, contracting companies, and workers, which angered the Turks but did not greatly reduce the scope of Turkish projects in Libya.

Politically, Ankara was finding it increasingly difficult to ignore the mounting evidence of Libyan involvement in international terror. That was a particularly sore point with the Turks, who had often been victims of both Armenian and Kurdish terror attacks. Nevertheless, economic interests still outweighed political ones in the 1980s, and Turkey did not take an unequivocal position against Libyan-inspired terror in the international arena.

As tensions mounted between the United States and Libya during the 1980s, official Turkey had to respond often to various incidents. In 1981, when two Libyan jet fighters were shot down by U.S. aircraft, Qaddafi called on Turkey to close down U.S. air bases in its territory. Ankara responded that there were no U.S. bases in Turkey (as distinct from NATO ones) and that because Turkey would not interfere in the defense considerations of other nations, they ought to refrain from interfering in Turkey's. Unfazed, Libya's president declared that his forces were capable of raiding U.S. bases in Turkey, Greece, and Italy. The Turkish Foreign Office chose to dub his statement "rhetoric which needs not be taken seriously," leaving the task of severely criticizing it to the Turkish press.[54]

The sharp contrast between Turkey's and Libya's positions on international terror had to be glossed over once again in April 1986, when the Americans bombed Qaddafi's offices in Tripoli.[55] Again Turkey's economic interests, coupled this time with the Arab world rallying on Libya's side, outweighed the Turkish position against international terror. Turkey only expressed the pious hope that U.S.-Libyan relations would not deteriorate further. That restraint, however, was not duly rewarded; in mid-1986 two Libyan citizens were arrested in Ankara and charged with an attempt to bomb the U.S. officers' club there. Several Libyan diplomats had been involved, but they were not brought to trial because of their immunity.[56]

During the 1990s Turkey's relations with Libya took on a totally different complexion. With oil becoming a marginal factor in political relations and the Middle East peace process gaining momentum, Turkey turned away from Libya and improved its relations with Israel. Libya was an international outcast during most of the decade as a result of the international sanctions imposed on it because of its involvement in the Lockerbie tragedy.[57] A rift between Turkey and Libya thus became unavoidable.

Even when pro-Islamic Necmettin Erbakan became prime minister (1996–1997, discussed later), reconciliation proved impossible despite his strenuous efforts. He was unable even to make the Libyans repay their huge debt to Turkish construction companies. In just one example of Middle Eastern inconstancy, it took less than two decades for Libya to be replaced by its sworn enemy—Israel—as Turkey's closest economic and strategic ally in the region.

Conclusion

The particular way in which some Turkish leaders viewed the 1973 energy crisis and its aftermath has had a significant impact on Ankara's political relations with the Middle Eastern oil producers, singly and severally. Those relations were already changing as early as mid-1974. In the main, the change consisted of an increasing Turkish willingness to reach political and strategic understandings—amounting in some cases to bilateral agreements—with as many oil exporters as possible. Despite the huge gap between the positions of Bülent Ecevit and Süleyman Demirel vis-à-vis the oil producers in particular and the Third World in general, in the late 1970s there was little difference in their approaches to the bilateral relations between Turkey and each of the oil producers.

Until 1973 Turkey had experienced no difficulty in either obtaining or paying for oil, and it had been able to deal with its Muslim neighbors according to its own political and strategic agenda. Economics as a whole, including energy, had not played a major role in Turkey's foreign policy considerations. That situation changed abruptly in late 1973, when the price of oil and the struggle to obtain it suddenly came to the fore. In early 1974 the Turkish Foreign Office developed a policy of reaching bilateral understandings with the Muslim oil producers, with the aim of obtaining oil under preferential terms but without making any political sacrifices, especially in Turkey's relations with the West. For a brief period it did seem as though Turkey would be able to maintain this policy of separating the oil issue from political relationships.

The standard-bearers of this separation policy were the Justice Party, as well as military and foreign policy circles in Ankara. From the start, however, they encountered strong objections from Erbakan's Milli Selamet Partisi (National Salvation Party) and many leaders of the CHP, who sought to use the oil issue as a lever to bring Turkey politically closer to the Muslim world. These opposing views clashed openly in early 1974 over the effort to make Saudi Arabia a major supplier for Turkey, when both Lütfi Doğan and Erbakan visited that country and came back empty-handed. It should be noted that Erbakan was under strict instructions, reflecting the supremacy of the separation policy at the time, not to make any political concessions (see Chapter 2).

Turkey's attempts to maintain a separation between oil and political and cultural affairs, however, made progress in its relations with the oil producers awkward and difficult. Thus, for example, when Turkey sent officials from the Foreign and Energy Offices to Baghdad to discuss oil instead of sending political leaders to discuss broader questions as well, the Iraqis were displeased.[58] The complicated issue of Turkish oil imports from Iraq and the pipeline project, it should be recalled, was solved only when Turkish prime minister Ecevit met with Iraqi vice president Hussein. It was only at that high level that a breakthrough became possible, and indeed, Hussein explained that talks had been deadlocked until then because they had been handled at the technical level.

Iraq, like all other oil producers, sought to take advantage of its new power as soon as the energy crisis started. Its objectives were crystal clear: to gain a greater share of Euphrates water and collaboration against the Kurdish threat in exchange for oil. By making Turkey sign a comprehensive agreement on all three issues at once, the Iraqis in effect forced the Turks to make political and strategic concessions, running against the grain of their crumbling separation policy.

The agreements reached between Turkey and Libya in January 1975 placed oil in a much wider context. This time even the Justice

Party, which only six months earlier had managed to block political discussions with Saudi Arabia, did not protest the comprehensive nature of the agreements with Libya and in effect acquiesced with that country's political demands. Moreover, Turkish public opinion was enthusiastic about the agreements, mindful of how both Saudi Arabia and Iran had dismissed Turkey's requests for preferential treatment and of how Libya had supported Turkey on the Cyprus issue.

Under Prime Minister Demirel, Turkey's political attitude toward its two chief oil suppliers, Iraq and Libya, changed only slightly. Demirel, wishing to maintain Turkey's pro-Western posture, included in his cabinet men like İhsan Sabri Çağlayangil and Kamran İnan, who in 1974 had led the resistance to Erbakan and everything he stood for. True, the "special atmosphere" Ecevit and Erbakan tried to create in Turkey's relations with the Muslim oil producers became awkward starting in April 1975. Demirel and Çağlayangil tried to implement the Iraqi and Libyan agreements without deviating from Mustafa Kemal Atatürk's legacy in terms of its pro-Western orientation and secular worldview. Relations with those two nations, which were then closer to the USSR than to the West, remained good and productive, but they lacked the romantic aura Ecevit had wished for in 1974.

Demirel's attitude during the years 1975–1977 was based on a rational approach, putting at the fore Turkey's immediate national interests, both economic and political. In this he differed from Ecevit, who had based his policies on ideological and emotional motives as well, believing as he did in the existence of long-term "common principles and interests" for Turkey and the oil producers. With Ecevit, history and philosophy had played as central a role as Turkey's immediate needs in shaping his policies. He believed short-term interests should not shape overall foreign policy and considered Turkey's relations with the various oil producers in terms of their contribution not only to Turkey or even to both parties but also to the well-being of the region, the Muslim world, and humankind as a whole.

Despite these differences in attitude, Turkey's policies toward the Muslim oil producers barely changed during the period 1975–1977. Turkey sought to develop economic and political ties with those producers that gave it preferential terms in oil purchases, flatly refusing to import oil from Muslim countries under normal market terms. There was a clear economic motive for this policy, but Turkey also firmly believed that as a member of the Muslim world, close both geographically and spiritually to the oil producers, it was entitled to special status as a customer. The businesslike attitude of Saudi Arabia and Iran in their responses to Turkish requests along those lines was interpreted in Ankara as patronizing and alienating, whereas the more politic approach taken by Iraq and Libya was regarded as exemplifying true friendship in time of need.

As long as Iraq and Libya continued to display such generosity, those in Turkey's political system who wanted stronger ties with Saudi Arabia and Iran had a hard time converting decisionmakers and public opinion to their views. The balance sheet spoke for itself: in the years 1974–1977 Turkey relied on Iraq and Libya, the less palatable choice politically (at least for Demirel and his supporters), because those nations together provided what Turkey needed in terms of energy supplies. Curiously, then, a coalition government led by the pro-Western Justice Party was conducting simultaneously a policy of moving ever closer to those OPEC members whose global orientation was the farthest from Turkey's. At the same time, relations with the pro-Western Saudi Arabia and Iran, two nations that should have been Turkey's closest regional allies, were superficial and perfunctory. Inevitably, then, Ankara courted the country that would give Turkey the best terms for oil and considered little else. Once Turkey's foreign exchange reserves ran out, leaving it unable to pay for its oil, a complete realignment of its foreign policy became necessary.

The series of barter agreements made by the Ecevit government during 1978 intensified bilateral relations with many Muslim oil producers. Turkey's new coalition government no longer regarded economic assistance and special terms for oil purchases as a precondition for good relations with those nations. The shift of public opinion in their favor was now part of the reality of Turkish politics. Thus at the CHP conference in July 1979, thousands of delegates rose to their feet and shouted "Libya! Libya!" for three long minutes, even though just before this display the Libyans had turned down Ecevit's request to increase the amount of oil Turkey was importing from them and the extent of both financial aid and trade with that country was disappointing.

The irrelevance of the differences in ideology between Demirel and Ecevit became apparent again in late 1977, when relations with Turkey's two chief oil suppliers—Iraq and Libya—began to deteriorate as a result of Turkey's inability to pay its debts. Toward the end of his term in office, Demirel realized it was no longer possible to base Turkey's relations with the oil producers exclusively on their willingness to grant his country preferential treatment in oil purchases. He concluded that Turkey should move closer to additional oil producers and expand the foundations of bilateral relations with each—a page out of Ecevit's book. Now the Justice Party was talking about "mutual interests" (or "balance") in Turkey's relations with the oil producers, whereas Ecevit had been talking all along about "common interests." Thus the two opposing leaders had maintained their differences in their attitude toward the oil producers as a group, but bilaterally their philosophies were virtually identical.

As we saw, the lessons of late 1977 made Turkey expand economic ties with the Muslim oil producers above and beyond oil. Labor force exports, the operation of Turkish contracting companies in the oil countries, and exportation of agricultural and industrial products were assuming greater weight in Turkey's bilateral relations with the various oil producers.

A similar solution called *bilateralism,* controversial though it was, was adopted by some other Western oil importers as well. Although the United States called for a joint struggle among oil importers against the producers, nations like France preferred to increase cooperation between each importer and its exporters. Turkey chose the latter way.

For most oil importers, particularly in the developing world, bilateralism was the inevitable solution. For Turkey, the days of debate over the need to develop bilateral ties with the oil producers were long past. Observing the increase in the producers' economic and political power, Turkey chose to rely on them for relieving the heavy burden of oil imports and protecting against possible future oil embargos. Even those who until 1977 had called for the separation of Turkey's economic problems from its foreign relations now had to recognize that economic damage control was at the top of the agenda and that everything else must follow suit.

The development of bilateralism in Turkey's relations with its oil producers culminated in the 1978 barter agreements, which were only partially successful: they improved Turkey's balance of trade but could not defuse the economically based social crisis that hit Turkey in 1980 or the political consequences of that crisis. But during the 1980s, when the foundations of Turkey's bilateral relations with its oil suppliers had shifted and Turkey began its dramatic economic recovery (see Chapter 5), politics again followed suit. By the end of that decade, political and strategic considerations resumed their central role in Turkish foreign policy decisionmaking.

This was first and foremost the result of a gigantic leap forward by

Turkey's economy. It has become modernized and globalized, pulling itself to an entirely new technological level. Now Turkey is looking for know-how and that can be found only in the West. In a sense, it can be said that Turkish foreign policy continues to serve the needs of the economy. But since the economy itself has changed, foreign policy has followed suit: now it is seeking to move Turkey closer to the European Union and to Israel as the only nation in the Middle East that can significantly contribute the new technologies Turkey requires. These trends are further explored in Part 3.

Notes

1. Syria had plans of its own, including construction of the al-Tabqa Dam on the Euphrates. When its reservoir was filled between 1974 and 1976, the flow of water into Iraq was reduced by half. This hurt Iraq both economically and politically, having caused unrest among Iraqi peasants in the southern part of the country.

2. *Middle East Economic Survey,* 22 January 1990.

3. In fact, that decision, which would cost Turkey billions of dollars, was made even before the United Nations officially imposed its sanctions on Iraq.

4. Bruce Kuniholm, "Turkey and the West," *Foreign Affairs* 70, no. 2 (spring 1991); Nuzhat Kandemir, Turkish Ambassador to the United States, in a speech, 19 March 1992 in Atlanta.

5. Quoted *Radical* (newspaper), 13 February 1999.

6. Except for short intervals, during which the United Nations allowed the use of the pipelines under its "food for oil" deal with Iraq, the pipelines remained dry most of the decade.

7. The CENTO alliance was established in 1959 to replace the defunct Baghdad Pact following the revolution in Iraq. Sponsored by the United States and Britain, it included Iran, Pakistan, and Turkey—the northern tier separating the Middle East from the Soviet Union.

8. The RCD was established in 1964 as an economic counterpart to CENTO and included the same members. It has never been much of a success because of fundamental conflicts of economic interest among its members.

9. See Christopher Rand, *Making Democracy Safe for Oil* (Boston: Little, Brown, 1975), pp. 160–164.

10. The shah told *Milliyet* (interview, 3 October 1973) that the pipeline would reach İskenderun, where the oil would be loaded on tankers or possibly be extended by another pipeline to Europe. On 24 July 1974, during an official visit to Paris, the shah reaffirmed this point (*Milliyet,* 27 July 1974).

11. *Pulse,* 24 May 1974.

12. *Günaydin, Cumhuriyet,* 9 July 1974.

13. The weekly *Briefing* wrote on 29 January 1975 that Turkey had lost all

hope of having an Iranian gas pipeline, because the shah had already approved an alternative plan to ship gas to Europe and the United States in tankers.

14. *Pulse,* editorial, 27 February 1975.

15. *Cumhuriyet,* 27 January 1975.

16. Interview with the author, Ankara, 9 May 1985.

17. *Hürriyet,* 8 June 1975.

18. See, for instance, op-ed section in *Cumhuriyet,* 9 June 1975.

19. Quoted in *Pulse,* 24 January 1980.

20. *Milliyet* interpreted this to mean either selling Turkey oil under preferential terms or depositing Iranian financial reserves in Turkish rather than Western banks.

21. *Pulse,* 19 February 1980.

22. *Hürriyet,* 11 April 1980.

23. *Tercüman, Hürriyet,* 11 April 1980.

24. *Milliyet,* 17 April 1980.

25. See *Milliyet,* 25 April 1980; *Cumhuriyet,* 30 April 1980.

26. *Hürriyet,* 23 April 1980.

27. I. Rabinovich and H. Shaked, eds., *Middle East Contemporary Survey* (Tel Aviv: Tel Aviv University, 1983–1984), p. 745.

28. A. H. Cordesman and A. R. Wagner, *The Iran-Iraq War* (Boulder: Westview, 1990).

29. Rabinovich and Shaked, p. 860.

30. Ibid., p. 654.

31. Ibid., p. 746.

32. Ibid., p. 674. Ten years later Turkish public opinion could afford a stronger reaction when Iran's leader, 'Ali Akbar Hashemi Rafsanjani, refused to pay tribute at the Atatürk mausoleum during a visit to Turkey. See Chapter 9.

33. The populations of five of the six republics (a total of 60 million people) speak languages that can be regarded as Turkish dialects, and most see themselves as ethnic Turks.

34. See *HaAretz* (Hebrew), 16 January 1992.

35. As stated by Foreign Minister Ismail Cem, *Turkish Daily News,* 23 May 2000.

36. See Aryeh Shmuelevitz, *Turkey's Experiment in Islamic Government, 1996–1997* (Tel Aviv: Moshe Dayan Center, Tel Aviv University, May 1999), p. 19.

37. Ibid., p. 26. At the time, that demand had been kept secret, but it subsequently found its way to the press.

38. In the Muslim world several organizations have the same name, which in Arabic means *God's Party.*

39. *Milliet,* 28 July 1999.

40. *Günaydin,* 23 August 1974.

41. Egypt, Libya's traditional rival, had taken a strong position in favor of the Greek Cypriots, becoming the Muslim nation that objected most fiercely to Turkey's intervention.

42. See *Pulse,* 6 September 1974.

43. *Tercüman,* 30 December 1974.

44. *Pulse,* 3 January 1975.

45. *Hürriyet,* 6 January 1975.

46. *Cumhuriyet* claimed later (8 September 1975) that there were also military agreements, particularly for cooperation between the two air forces.

47. *Pulse,* 6 January 1975.

48. *Cumhuriyet,* 12 January 1975. Jalud said this during the Friday service at an İstanbul mosque. He added that any military attack on Turkey would now constitute an attack on Libya, in which case all of Libya's armed forces would be put at Turkey's disposal.

49. *Pulse,* 17 January 1975.

50. *Hürriyet,* 6 March 1975.

51. *Pulse,* 13 August 1975.

52. In an article published in 1982, Professor Mükerrem Hiç expressed his opinion that Turkey had developed economic ties with the radical Libya and Iraq rather than with the more moderate Muslim nations not because of political proclivities but because those two nations imposed themselves on Turkey: Libya because of Qaddafi's revolutionary philosophy, and Iraq because of its need for an outlet to the Mediterranean. See his "Economic Policies Pursued by Turkey," *Orient* 23, no. 2 (June 1982): 220–242.

53. *Hürriyet,* 30 July 1978; *Pulse,* 31 July 1978.

54. I. Rabinovich and H. Shaked, eds., *Middle East Contemporary Survey* (Tel Aviv: Tel Aviv University, 1980–1981), p. 855.

55. The raid took place on 15 April, following the bombing of a discotheque in West Berlin in which a Turkish woman, among others, had been killed. Although Qaddafi was unharmed, fifteen Libyans died in the Tripoli raid.

56. Rabinovich and Shaked, p. 653.

57. Two Libyan officials were arrested for blowing up the TWA flight over Lockerbie, Scotland. One of them was found guilty in December 2000.

58. Similar frustration was expressed at the time by both the Saudi ambassador in Ankara (*Yankı,* 1 March 1974) and the Iranian ambassador (*Hürriyet,* 24 December 1974).

9

Turkey and the Arab-Israeli Conflict

Ever since the establishment of the State of Israel in 1948, Turkey had been vacillating endlessly about the nature of its relationships with Israel on the one hand and the Arab countries on the other. As a Muslim nation it had centuries of cultural and religious tradition linking it to the Arab world. Also, its geopolitical position, as noted extensively earlier, required it to carefully consider the Arabs' interests. Yet the tenets of Kemalism—which emphasized a Western orientation, secularization, and modernization—have tended to produce much stress in Turkey's relations with many Arab nations, including its closest neighbors, Syria and Iraq. Turkey has differed from the Arab world on a series of issues in which it has found itself much closer to Israel. Both nations are democracies, unlike the entire Arab world; and along with Iran they are among the few non-Arab states in the entire region. In its earliest years Israel had been regarded by many nations, including Turkey, as leaning toward the East (mainly because it received arms during its war of independence from the Soviet bloc and its leaders had a socialist background), but since the early 1950s Israel had been following a clearly pro-Western line, much like Turkey, whereas more and more Arab countries were becoming pro-Soviet. The Turks also remembered well the 1916 Arab revolt against the Ottoman Empire, which Turkish collective memory considered a stab in the back. All of these factors played a role in Turkish foreign policy regarding the Middle East conflict in the 1950s and 1960s.

Besides these long-term considerations, Turkish policy was also shaped by immediate reactions to current events. For instance, Turkish public opinion was fairly supportive of Israel during the early years. The military and technological capabilities developed by that young nation within a short period, particularly its victory over the combined military forces of many Arab nations, impressed the Turks (many of whom gloated over the defeat of the Arabs). And the Turks have also tended to attach great importance to the influence of world Jewry as an international supporter of Israel. The Turks, like many others world-wide, have regarded American Jews in particular as enjoying enormous power and influence over U.S. public opinion and foreign policy. Additionally, Turkey's relations with its Arab neighbors have been uneven against a background of unremitting bilateral and multilateral conflicts, including Syrian territorial demands for the İskenderun-Alexandretta area.

The Arabs, for their part, have exerted constant pressure on Turkey since the early 1950s in an ongoing effort to win it over to their side of the conflict. They could hardly ignore the fact that Turkey was the only Muslim nation in the world that not only recognized Israel in the first place but developed extensive relationships with that country. As a matter of routine, each news item about yet another move to bring Ankara and Jerusalem closer raised a storm of protests by Arab representatives in Turkey. Yet those protests have had little effect—which has been a source of permanent frustration for the Arab world. The more fundamentalist Arab countries have put together Turkey's alienation from Islam and its favorable attitude toward the Jewish state and have been waiting to bring their own pressures to bear on Turkey.

Israel has developed a system of counterpressures on Turkey, albeit in a more subtle way. Besides the significance Jerusalem attached to the very existence of its relations with Turkey, it was also concerned about any move that might bring Turkey closer to one Arab state or another in the military or strategic sphere (as happened when the Baghdad Pact was created in 1955). Any increase in the military capabilities of any Arab nation was a cause for worry in Jerusalem. Israel's efforts took place in two parallel channels: the normal, diplomatic channel and also through Washington. U.S. Jewish leaders have been mindful of Turkey's importance for Israel and have made a point of meeting with each Turkish leader who has visited the United States. The Turks, for reasons noted earlier, have always been obliging.

Such meetings have developed a routine. The Jewish leaders talk about the importance the West attaches to the stability of links between the only two democracies in the Middle East. The Turks point out the liberties the Jewish community enjoys in their country (which have been extensive over the past 500 years), mention that Turkey has been the only Muslim nation to recognize Israel, and ask for help in improving Turkey's image in the United States and in persuading Congress of the need to provide foreign aid. The Jewish leaders have usually drawn the line when the Turks have asked them to help thwart the debate taking place intermittently in the United States over commemorating Armenian victims during World War I. Israel's official representatives in Washington have always kept their distance from such meetings.

Ankara has been able to strike a balance between its interest in maintaining good relations with both Israel and the Arab world and the pressures and counterpressures brought to bear on it by keeping a low profile in its relations with Israel. This was true even after 1958, a major turning point in Turkish-Israeli relations, as explained later. Following the Iraq revolution and the growing strength of the Nasserist movement, the Turkish prime minister at the time, Adnan Menderes, concluded that the chances of the Arab world becoming pro-Western were nil and that Turkey should therefore move closer to Gamal 'Abd-al-Nasser's greatest enemy, Israel. Thus began a process that resulted in a strategic agreement between Turkey and Israel.

In retrospect, Turkey's position on the Middle East conflict until the mid-1960s was fairly comfortable for Israel. Bilateral relations were stable, whereas Turkey's relations with the Arab nations (except for Iraq for a few years) were rather vague, despite perfunctory Turkish periodic declarations of support for the Arab cause. Things began to change in the mid-1960s, first because Turkey's pro-Western attitude was weakening and second because Turkey found itself in need of support at the United Nations over the Cyprus issue. Since Israel had annexed the eastern part of Jerusalem immediately after the Six Day War in 1967, Turkey was finding it more and more difficult to maintain its relative neutrality toward the Arab-Israeli conflict. Turkey now emphasized its support of the Arab cause, calling for Israeli withdrawal from all Occupied Territories and for full self-determination for the Palestinian people. Without breaking off diplomatic relations with Israel, Turkey was increasingly bringing its position in line with the Arab side, both bilaterally and in international and Islamic forums.

From the mid-1960s to the mid-1970s the development of the Arab-Israeli conflict, rather than the Cold War, guided Ankara's attitude toward Israel. Even though most Arab countries remained anti-Western, that was no longer an obstacle for Turkey. In 1948 Israel's military victories had won it Turkish admiration, but the 1967 victory had the opposite effect. By annexing East Jerusalem Israel went too far as far as Turkey was concerned. From the mid-1970s through the 1980s, as already discussed, Turkey moved closer and closer to the Arab position in the Middle East conflict. The traditional reasons for maintaining ties with Israel (Israeli democracy, perceived Israeli and Jewish influence in Washington, and so on) kept Turkey from severing diplomatic relations—the top Arab demand—but they were insufficient to give those relations substance until well into the 1980s.

The rapprochement with several oil-producing countries, the penetration of Turkish contracting companies into the markets of many Middle Eastern countries, and growing Islamization domestically have made Turkey far more responsive to Muslim sensitivities, particularly with regard to Jerusalem's holy places. Such events as the fire at the al-Aqsa Mosque, the Jerusalem Law, the Lebanon War—particularly the massacre in Palestinian refugee camps near Sidon in 1982—the bombing of Beirut, and, later, the Palestinian uprising (intifada) have all had a significant impact on Turkish-Israeli relations. Turkey's sensitivity to both the Jerusalem issue and Israel's treatment of Muslim and other Arab civil populations in Lebanon and the Occupied Territories remained the same even when it was no longer at the mercy of Arab oil producers. It has also continued to maintain good relations with the Palestine Liberation Organization (PLO).

As we have seen, the domestic crisis in Turkey, which reached its peak in 1980, proved fertile ground for a combined Arab pressure—economic, political, and religious-cultural. A weakened Ankara had to give in on all counts: relations with Israel were downgraded, and Saudi Arabia and other Muslim nations were allowed to sponsor Islamic activities inside Turkey and among Turks abroad. But after a decade of political stability and economic recuperation during the 1980s, Turkey was less vulnerable to pressures in the 1990s. It also used that period to improve its bargaining position vis-à-vis the Arab countries, controlling the flow of water into Syria and Iraq and becoming a net lender to Iraq, Iran, and Libya. Its political attitude toward the Arab-Israeli conflict (as distinct from its humanitarian and religious concerns) changed: in January 1992, shortly after the Arab-Israeli peace conference in Madrid, Turkey felt sufficiently confident to resume full diplomatic relations with Israel at the ambassadorial level.

Endless Friction with Syria

Among Turkey's Muslim neighbors, its relations with Syria have always been the most problematic because of the many points of friction between the two nations. Remnants from the Ottoman period resurfaced as soon as Syria became independent in 1943. For many decades since, the main bone of contention between the two nations was the Alexandretta area (now the Turkish province of Hatay), where the border between the two countries reaches the Mediterranean. In the 1970s and 1980s, however, two additional problems became more acute— Syria's share of Euphrates water and its support of anti-Turkish terrorism. During the 1980s Turkey's policy toward Syria was often characterized as one of "water for security."

Although included within the boundaries of the French mandate over Syria, the Alexandretta area was annexed to Turkey in 1939 with French consent. At the time, Syrian resentment was bitter but was directed mainly against France. Once independent, Syria redirected its demands toward Turkey, claiming the province was historically and geographically part of the Syrian homeland and that its population was largely Arab. For the Turks, it was just another Turkish province with an Arab minority. Syria has still not dropped its demand to reclaim Alexandretta, and Turkey has been unwilling to consider the demand.

When diplomatic relations were established in 1943, the Turks demanded that in return for their recognition Syria would give up any claim to Alexandretta. The Syrians refused, and it took lengthy mediation efforts by Iraq before the Syrians agreed in 1946 not to put up any *formal* demand for the area. Nevertheless, public opinion as well as the press continued to express the sentiment that Alexandretta was an indivisible part of Syria. In the 1950s and 1960s many anti-Turkish demonstrations and even a few border clashes took place in Syria. Later, the Syrians gave their 1946 agreement a narrower interpretation; for instance, they published official maps in which the area was shown as a Syrian territory. They also approached Turkey, seeking to open negotiations on the issue, but the Turks flatly rejected the overtures, warning Syria "not to play with fire," as Prime Minister Menderes put it in 1955. In this conflict Syria has enjoyed the support of several Arab countries, including Saudi Arabia; during the 1980s, for example, the Saudis refused to grant entry visas to residents of the area who wanted to go on pilgrimages to Mecca.[1]

There were other difficult points in the 1950s. Syria took a jaundiced view of the developing ties between Turkey and Israel, as well as of Turkey's efforts to consolidate a pro-Western regional alliance. When the Baghdad Pact was created in 1955 with Turkish and Iraqi participation, Syria began to gravitate toward Egypt, developing first a defense treaty and then full political union (the short-lived United Arab Republic) with that country. The bloc enjoyed Soviet military assistance, which Turkey, Israel, and the West as a whole tended to regard as a threat to stability in the region. In August 1957, when the Syrian chief of staff visited Moscow, Turkey concentrated its forces along the border. The USSR responded by warning Turkey not to interfere with Syria's affairs.

But in the 1960s, as noted earlier, Ankara began changing its policy orientation in a way that much better suited the Syrians. Against a background of disenchantment with the West (over Cyprus) and closer relations with the Soviet Union, Turkey and Syria were drawing closer. The shift was noticeable during the Six Day War, when Turkey announced it would not allow the use of North Atlantic Treaty Organization (NATO) bases in its territory to deliver supplies to Israel. It also informed the Syrians that it would not concentrate forces along the border and even sent Syria humanitarian aid—foodstuffs, clothing, and medical supplies. Soon thereafter, on 21 August 1967, Syrian foreign minister Ibrahim Mahous publicly expressed the appreciation and gratitude his nation felt toward the Turkish people. Turkey's attitude was the same six years later, during the Yom Kippur War, which again brought Turkey and Syria much closer.

At the same time, significant policy shifts were occurring in Damascus under Hafiz al-Assad. Diplomatic relations with Turkey were upgraded, and the two countries' foreign ministers, Haluk Bayülken and 'Abd-al-Halim Khaddam, exchanged visits in 1972–1973. They discussed commerce, air links, tourism, and cultural affairs, as well as political matters such as Cyprus and border problems (smuggling, unauthorized crossings, and untoward incidents). During that time a longstanding problem was finally resolved: a settlement was reached over nationalized property belonging to Syrian citizens in the Hatay province (Alexandretta) and to Turkish citizens in Syria.

In the late 1970s relations cooled again. Syria's involvement in the Lebanese civil war was a source of concern for Turkey, indicating as it did that Syria had not abandoned its "greater Syria" vision, which included Alexandretta. Also, Syria regularly supported the Greek position on Cyprus, not only at the United Nations but also at the Islamic

Conference. Most serious, however, were accumulating indications that Syria was training Turkish dissidents and irredentist elements in PLO camps in its territory. Several Turks caught in clashes with police told interrogators they had trained in Syria. The late 1970s witnessed an increasing wave of terror in Turkey proper, as well as assassinations of Turkish diplomats overseas; the terrorism continued well into the 1980s. Armenian terrorist groups, well connected to the PLO and receiving training in both Syria and Syrian-controlled parts of Lebanon, were responsible for the deaths of more than thirty Turkish diplomats (including family members) during this period, most in Europe and the United States. The Syrians kept denying any involvement, but information from various sources made those disclaimers difficult to believe.

In the early 1980s, however, following the upsurge of Turkish trade with the Arab world—including Syria—high-level meetings resumed. Both the foreign ministers and deputy prime ministers of the two nations exchanged visits in the years 1981–1983, discussing both trade and terror. When Deputy Prime Minister Turgut Özal visited Damascus in March 1982, a series of economic and commercial agreements were signed, although they failed to bring about a significant change in the patterns of trade between the two nations. Syrian exports to Turkey, which had increased significantly in the late 1970s (reaching $129.5 million in 1981, up from $29 million in 1977), dropped again, plunging to $5 million a year during most of the 1980s and 1990s. Turkey's exports to Syria stabilized at about $60 million a year.[2]

In the late 1980s and 1990s the water problem entered the equation, combining with the unsolved terrorism problem to cast a heavy shadow over Turkey's relations with Syria. The Syrians, claimed the Turks, were harboring not only Armenian but also Kurdish terrorists, allowing the Parti-ye Kerkaran-î Kürdistan (PKK, Kurdistan Workers' Party) to train in camps inside Syria and in Lebanon. Syria's claim that it did not control Lebanon's Beka'a Valley, where most of the camps were located, was ridiculed by the Turkish press.[3] Even when Syria was finally pressured into deporting the Kurdish leader Abdullah Öcalan, as noted earlier, that move only discredited its previous disclaimers. Turkish public opinion was outraged and demanded that the government take a firmer posture against Syria, not only in terms of water supplies but also that it launch military strikes against Kurdish training camps in Syria and Lebanon.[4]

During the 1980s the insurgency increased along Turkey's eastern border, in both scope and severity. Kurdish separatists were largely responsible. Infiltrating from Syria, Iraq, or Iran, they constantly threatened lives and property, including land transportation from Turkey to those countries and the Iraqi pipeline. According to Turkish Army announcements, the number of casualties among Turkish security forces and civilians during this period was about 2,000 (almost 600 in 1988 and 1989 alone). Around 1,500 Kurdish separatists were also killed, 60,000 suspects were arrested, and more than 2,000 were convicted. In a single attack on the town of Ikiyaka on 24 November 1989, terrorists massacred twenty-nine people, including thirteen children, then fled across the border. The figures climbed rapidly during the 1990s and the Turkish official estimate of the overall number of casualties in the Turkish-PKK clashes between 1984 and 1999 is around 30,000 dead on both sides.

The Turks have understandably been sensitive to any sign of support for Kurdish separatism. Having accumulated sufficient evidence that Syria was actively aiding and abetting both Kurdish and Armenian terrorists, they have held Damascus responsible for unrest along Turkey's border and increased displays of Kurdish nationalism domestically.

Syria and Euphrates Water

A long-standing problem in Turkey's relationship with Syria, as noted earlier, has been the allocation of Euphrates River water. Flowing into Syria from Turkey and then on into Iraq, the river carries 80 percent of Syria's total water potential. Since the late 1980s Turkey has been engaged in a huge development effort along the headwaters of the Euphrates, estimated at a total of $56 billion by 2005. Inevitably, that effort has reduced the amount of water reaching Syria, which has little prospect of developing alternative water resources. Syria, like Iraq, has been complaining all along that Turkey disregarded existing agreements on allocation (agreements whose existence was challenged by Turkey), and such complaints had an effect internationally. As a result, Turkey was having a difficult time raising external financing for those projects. Nevertheless, Turkey was persistent, work went on, and tensions kept mounting.

On 13 January 1990 Turkey began filling the Atatürk reservoir, the centerpiece of the entire development plan. For an entire month, Euphrates flow into Syria was one-fifth its regular capacity. Although

Turkey gave both Syria and Iraq advance notice of its intentions and had increased the flow previously, the damage to the Syrian economy was extensive. Both agriculture and power supplies were hurt, and with power cuts came secondary damages to industry.[5]

In fact, there has never been a formal agreement between Turkey and Syria over the regulation of water allocation. There is a joint water committee, and a protocol was signed during a visit by Turkey's prime minister to Damascus in July 1987, whereby Turkey undertook to allow a flow of 500 cubic meters per second even while the Atatürk reservoir was being filled. In the absence of clear-cut international law on the matter, Syria resorted to commonly accepted norms formalized in the Helsinki Rules document published in 1966 by the International Law Association, declaring that all riparian nations are entitled to an equal share of any international river's water. The document states that if an upstream country decides to change the rate of flow, it should inform the downstream countries—and receive their consent. The Turks did notify their neighbors of their intentions but received no consent from either Syria or Iraq.

Against this background, in January 1990 Syria launched a diplomatic campaign against Turkey. It called on other Arab nations to intervene, and the Arab League demanded at once that the flow of Euphrates water be restored to its normal rate. Turkey sent its foreign minister on a tour of Arab capitals, seeking to convince the Syrians and others that it had no intention of using water as a political weapon.

Relations during the 1990s still reflected the trauma of the January 1990 stoppage. Syria, aware of Turkey's technical ability to close the faucet, demanded that no recurrence take place (and so far none has). The Syrians also demanded that water flow be regularized, and Turkey agreed to provide a certain amount of water annually, even if that amount fell short of Syrian demands. The Turks complained that Syria was making inefficient use of its water supplies and wasting precious Euphrates water.

Yet another cause for tension in Turkey's relations with Syria in the mid-1990s was the tightening of military links between Turkey and Israel. As the Syrians saw it, this military alliance was aimed directly at them. In May–June 1996, after Turkey and Israel had signed a series of military agreements, Syrian-Turkish relations deteriorated nearly to a point of armed conflict. The deterioration was halted when Necmettin Erbakan, the Arabs' staunchest supporter in Ankara, became prime minister. But he was removed from office in July 1997, and Syrian support of PKK Kurdish separatists became the main issue in relations between the two countries—overshadowing not only Turkey's relations with Israel but also the water problem, even though all these issues remained tacitly linked, at least for Turkey.

In May 1998 Turkey issued several official warnings to Syria regarding its support of the PKK. Subsequently, those warnings gave way to threats. As the commander of the Turkish army, General Atilla Ates, put it, "We do not have any patience left. . . . The Turkish nation will have to take every possible step."[6] At a meeting of the National Security Council on 30 September 1998, the chief of staff, General Huseyin Kivukoğlu, said that "a state of undeclared war with Syria" already existed, although President Süleyman Demirel was slightly more guarded, noting only that Turkey's patience with Syria was "running thin."[7]

Belligerent statements by top-ranking Turkish officials continued in early October, and the newspapers were running wild on 7 October 1998: "Either APO [Öcalan] or war" (*Milliet*); "Turkey will run over Syria" and "be ready for war" (*Sabah*); "we will enter from one side and leave through the other" (*Hurriyet*). A true crisis was in the air, and the international community became worried. Egypt and Iran quickly offered their good offices as mediators, and by 20 October 1998 an agreement had been concluded between Turkey and Syria, amounting in effect to a complete surrender by Syria. The next day Prime Minister Mesut Yılmaz announced that Öcalan was no longer in Syria; he was now in Moscow.[8]

The air was cleared almost overnight. Both parties to the agreement were well aware that they faced more important issues. Economic, tourism, and trade relations were resumed. A series of mutual high-level visits began, including several military delegations. And the water issue came back to the fore. Said Foreign Minister Ismail Cem, "We shall not leave Syria without water."[9] In June 2000, when Turkey's new president, Necdet Sezer, represented his nation at President Assad's funeral, the Turkish press hailed it as heralding a "new chapter" in Turkey's relations with Syria.

Turkey and Egypt

Until 1914, Egypt was nominally part of the Ottoman Empire, but it was able to conduct foreign policy at variance with the High Porte's. When it became an independent kingdom, relations with the former metropolitan power were minimized and later became rather hostile. The Egyptian revolution made Egypt turn its back on the West and move closer to the Soviet Union, as well as declare itself leader of the Arab struggle against Israel—much to Turkey's chagrin. During the late 1950s, then, these nations' foreign policies were diametrically opposite: Turkey was a member of NATO, closely tied to the United States, and friendly to Israel, whereas Egypt was linked militarily to the East, hostile to the West, and adamant in its enmity toward Israel.

Turkey's efforts at the time to establish a regional defense system were the main cause of contention between the two countries. Besides the Cold War implications, Egypt regarded those efforts as an attempt by Turkey to acquire regional hegemony, a status it had been seeking for itself. When the United Arab Republic was created in February 1958, relations deteriorated even further. Egypt joined Syria in its claim for the Alexandretta area—an issue that, as noted earlier, cast a heavy shadow over Turkey's relations with Syria at the time. Furthermore, Egypt also supported the Greeks on the Cyprus issue. As a result, relations between Egypt and Turkey were tense until the mid-1960s.

At that time Turkish foreign policy began to change, with a view of moving closer to the Eastern bloc, the Third World, and the Arab countries. This change brought about a perceptible thaw in its relations with Egypt as well, even though the gap was still wide because Egypt could readily interpret the change as a move to enhance Turkey's claim to hegemony in the Middle East. Still, Egypt welcomed Turkey's refusal to allow the use of NATO bases in its territory for shipments to Israel during the Six Day War, as noted earlier.

Turkey's bilateral relations with Egypt began to improve, slowly but surely, only after Egyptian president Nasser died and Anwar Sadat replaced him in late 1970. Sadat steered Egypt on a more pro-Western course and showed willingness to resolve outstanding problems between his country and Turkey. In fact, Ankara had several occasions to congratulate Sadat for his "extraordinary" and "responsible" statesmanship.

It could have been expected, then, that Sadat's decision to go to Jerusalem in 1977 would had been welcomed by the Turks, who had repeatedly called on him to show greater moderation in the Arab-Israeli conflict. But Turkey's reaction was cautious, reflecting yet another change in its foreign policy in the wake of the 1973 energy crisis. At the time of Sadat's visit, Turkey was just beginning to feel the impact of the oil price hikes, as discussed earlier; now it could not afford to alienate those Arab oil producers, such as Libya and Iraq, who took strong exception to Sadat's initiative. Although the press and public opinion welcomed the development, on the whole official Ankara was much more reserved.

Turkey's growing dependence on Iraq and Libya caused it to reconsider a trip, planned well in advance, by Foreign Minister İhsan Sabri Çağlayangil to Cairo. Scheduled for December 1977, the trip would be the first official visit to Egypt by any foreign dignitary since Sadat's visit to Israel. Now Ankara was confused. The Egyptians, realizing this, announced their willingness to make a few gestures toward Turkey, including paying compensation for expropriated Turkish property, developing a more balanced policy on Cyprus, and appointing an Egyptian ambassador in Ankara—long overdue because the last one had left office nearly a year before.

Foreign Office officials tried to present Çağlayangil's visit as normal, intended to improve cooperation between Turkey and Egypt in various areas of mutual concern. They emphasized that it had nothing to do with Sadat's visit, which had been scheduled ahead of time. Postponing the trip, however, would be interpreted as taking sides in an internal Arab affair, which would run against the grain of traditional Turkish foreign policy.

All of this failed to impress the many Turkish political leaders who objected to Çağlayangil's visit. These included even some members of the ruling Justice Party, led (not surprisingly) by Energy Minister Kamran İnan, who argued that the visit could jeopardize Turkey's relations with other Muslim nations (first and foremost Iraq and Libya) and drag Ankara into an internal Arab affair. Several politicians, including Home Minister Oğuzhan Asiltürk and Labor Minister Fehmi Cumalıoğlu, demanded that the visit be redefined as a private one. Others, including Deputy Prime Ministers Erbakan and Alparslan Türkeş, demanded that the foreign minister be dismissed after he disregarded their objections and went on the official visit to Cairo as planned.[10]

The nations leading Arab opposition to the visit were headed by Iraq and Libya, who in late 1977 were in a good position to exert pressure on Turkey. The Libyans, never overly fond of Egypt, were opposed to Çağlayangil's visit even before the Sadat initiative. The Iraqis expressed objections to Deputy Prime Minister Erbakan when he visited in November 1977, and at the same time the Libyans made strong protests to another Turkish visitor, Deputy Prime Minister Türkeş.[11]

Still, no public criticism or threats were made, because Baghdad and Tripoli did not want to appear as interfering in Turkey's affairs. Nevertheless, their objections proved rather effective; Çağlayangil kept a low profile during his visit, took care to discuss only bilateral affairs, and called on both Mahmud Riad, the secretary-general of the Arab League (a leading opponent of Sadat's initiative), and the PLO representative in Egypt, Sa'id Kemal.[12]

The Turkish press speculated that Libya and Iraq might reduce their financial assistance to Turkey or even disrupt the flow of oil, but that did not happen.[13] They did succeed, however, in robbing the Çağlayangil visit of any regional significance: Turkey failed to support Egypt's president in his trailblazing move. Still, the fact that Turkey insisted on the visit taking place despite those pressures, which aimed at what at the time was a matter of survival—namely, oil supplies—indicated its adherence to its policy of keeping its diplomacy and its economic problems separate. As we saw earlier, that determination did not last much longer.

Ankara's relations with Cairo improved greatly in the early 1980s, especially following President Husni Mubarak's visit to Turkey in February 1985. That was the first time an Egyptian head of state had visited modern Turkey, a real breakthrough in view of a very long and troublesome history. In January 1986 President Fahri Korutürk made a return visit, and in February 1988 Prime Minister Özal went to Cairo at the head of a large delegation of Turkish businessmen. Following a series of economic agreements signed during those visits, the volume of trade between Turkey and Egypt began to expand—from $50 million a year in the early 1980s to $200 million by the end of the decade. Cultural relations and tourism grew apace, and cooperation in various economic areas increased.

Turkey's relations with Egypt have been strongly affected by the tightening links between Ankara and Jerusalem since 1993 (discussed in the next section). Egypt regarded itself threatened by what it considered as a Turco-Israeli military alliance and strongly protested the 1996 military agreement between Turkey and Israel. During the years 1997–1999 several high-ranking Turkish officials went to Cairo in an effort to reassure the Egyptians.

Relations improved considerably in 1999, following a successful intervention by Egyptian president Mubarak in the conflict between Turkey and Syria over the Turkish demand to deport Abdullah Öcalan from Damascus. The mediation efforts, in which the Iranians took part as well, resulted in the October 1998 agreement between Syria and Turkey, which satisfied the Turks. They thanked Mubarak profusely and immediately launched discussions on a series of economic agreements. The highlight of this renewed Turkish-Egyptian rapprochement concerned an underwater gas pipeline intended to provide Turkey with huge amounts of Egyptian natural gas.

Still, major differences exist between the two nations on issues relating to regional balance of power, as each nation has historically regarded itself as *the* regional power in the Middle East. Egypt is still suspicious of Turkey's military links with Israel, and the Turks remain unwilling to break away from Israel to placate Egypt and other Arab nations.

Turkey's Relations with Israel

In November 1947 Turkey voted against the UN resolution that led to the establishment of the State of Israel. But in March 1949, following a series of armistice agreements between Israel and the various Arab nations that participated in the 1948 war, Turkey officially recognized Israel. In late 1949 diplomatic relations were established at the ministerial level (that is, one level below the full or ambassadorial one). In justifying that move to the Arab countries, the Turks explained that an armistice agreement also amounted, in effect, to recognition.

Israel has attached great importance to its relations with Turkey, a Middle Eastern Muslim nation that could possibly serve as a bridge to the Arab world. Thus Eliahu Sasson, one of Israel's leading diplomats, was pulled out of his mission as head of the Israeli team negotiating with King Abdullah of Jordan to become Israel's first representative in Ankara.[14] Israel's leaders, some of whom (including Prime Minister David Ben-Gurion and Yitzhak Ben-Zvi, who soon became president) had been students in Istanbul before World War I, realized the strategic significance of those relationships and the numerous matters of mutual interest involved. Israeli leaders were also mindful of the large Jewish community in Turkey. In the past, Turkey had provided a safe haven for exiled Jews, from those expelled from Spain in the late fifteenth century to refugees escaping Europe before, during, and after the Holocaust. In 1948 around 100,000 Jews were living in Turkey. Many emigrated to Israel during the 1950s, but those remaining (about 25,000 at present) have enjoyed full civil liberties.

Turkey's relations with Israel developed rapidly during the 1950s. In July 1950 a trade agreement was concluded, and in February 1951 air links were established. Turkey and Israel coordinated their voting at the United Nations and exchanged numerous parliamentary delegations and visits by journalists. In 1953 a cultural agreement was signed, negotiations for upgrading the level of diplomatic representation were begun, military attachés were appointed in both legations (for Israel, this was only the fourth military attaché ever sent abroad), and an Israeli navy flotilla visited Istanbul.

Israel's Middle Eastern policy at the time was based on the idea that the non-Arab nations in the region should stick together, and that meant Turkey first and foremost. In September 1953 Israel's minister in Ankara, Morris Fisher, reported to Jerusalem that he had brought up the idea of cooperation in discussions with Foreign Office officials there. The time was ripe, he felt, following a visit to Ankara by the Pakistani chief of staff. Pakistan was eager to tighten its ties with Turkey, whereas Turkey was willing to move closer to Pakistan only if Iran would do so as well. Fisher thought Israel should also take part. As he wrote to his home office:

> It seems to me that since the Arab countries are not going to play a major role in the [pro-Western] defense arrangements for the northwestern part of Asia, Israel can have a more easy access to such nations as are friendly to Turkey and have no direct conflict with Israel, namely Persia and Pakistan. . . . Should nations like Iran or Pakistan become closer to Israel, the Arab countries will have to abandon their hopes for Israel's permanent isolation and their anti-Israel activities would be frustrated.[15]

But Turkey's foreign policy considerations made it turn the other way, preferring to tie its fate to the pro-Western Arab countries in the region, then led by Iraq. Thus when the Baghdad Pact was created in February 1955, it had an adverse impact on Turkey's relations with Israel. During negotiations, Turkish prime minister Adnan Menderes and his Iraqi counterpart, Nuri al-Sa'id, agreed, among other things, to coordinate their voting at the United Nations on all issues relating to the Middle East conflict. Consequently, several negotiations with Israel on further agreements were broken off, and Turkish leaders began making statements in support of the Arab cause against Israel. As the chargé in Ankara, Moshe Allon, explained in 1957, the Turks believed that "Turkey, as a NATO member, was the only nation capable of linking the

Baghdad Pact with NATO, gaining power and prestige in the region, and achieving regional hegemony with Western support."[16]

Following the Sinai-Suez campaign in late 1956, relations between the two countries deteriorated even further. The Turks recalled their minister from Tel Aviv, leaving at the head of their legation a chargé d'affaires ad interim—a situation that prevailed until 1992. In an official statement, the Turkish Foreign Office explained that the minister would resume office only when "the Palestine problem would be resolved in a satisfactory and lasting manner in accordance with UN resolutions."[17]

Israel's former minister in Ankara, Eliahu Sasson (then ambassador in Rome), was rushed to Ankara to ask Menderes for a better explanation. He was told that Turkey attached supreme importance to the Baghdad Pact and would do anything it took to strengthen it. In particular, Turkey was willing to pay any price to protect its ally Iraq from vicious allegations by other Arab nations, as though by collaborating with Turkey it was indirectly collaborating with Israel. Arab propaganda was also aimed at Turkey, accusing it of military collusion with Israel. And finally, Turkey felt it had to create proper conditions for such countries as Saudi Arabia and Lebanon to join the pact, directly or indirectly. For the creation and maintenance of this "buffer against Communist inroads," Turkey was willing to sacrifice its hitherto close relations with Israel.[18]

Subsequently, Turkish-Israeli relations deteriorated. Prime Minister Menderes neatly summed up his new position: "The Middle East is a common asset of the Muslim peoples, Turkish and Arab."[19] Menderes was the dominant figure in shaping Turkey's Middle Eastern policy, and the Israelis realized that any change—for better or worse—was up to him. As Allon wrote to Jerusalem (23 September 1957): "Menderes is the mover and shaker of Turkey's foreign policies, as well as domestic and economic policies. His associates have but limited influence on him. In order to achieve any change or turnabout in Turkey's policy vis-à-vis Israel, Menderes himself must be persuaded—through a direct contact with him. I do not believe such an attempt has any chance of success through indirect contacts, no matter who the intermediary would be."[20]

Thus a meeting was arranged in Paris on 15 December 1957 between Menderes and Sasson, who had become a personal friend of Menderes's during his term of office in Ankara a few years earlier. Allon proved right, because the change in Turkish policy became apparent immediately. Foreign Office officials later told the Israeli chargé in Ankara that as soon as the meeting was over, Menderes hastened to modify the draft of the address he had prepared for the NATO conference that had brought him to Paris in the first place on matters relating to Israel.

The Collapse of the Baghdad Pact

On 14 July 1958 a revolution shook Iraq. King Feisal II and his powerful prime minister, Nuri al-Sa'id, were assassinated, and General 'Abd al-Karim Kassim became ruler. His policy was markedly anti-Western, and the Baghdad Pact thus lost its mainstay. At the same time, factional strife in Lebanon was developing into a civil war, and the West decided to intervene. U.S. marines were sent there through the Turkish air base Incerlik, much to the dismay of many Arab countries. Turkey suddenly found itself at odds with most of the Arab world, and the Israelis realized that a new window of opportunity had opened for them.

The old idea of an "alliance of the periphery" was revived. Turkey was already moving closer to Israel, as explained in a "personal" letter sent by Moshe Sasson, a senior official in the Middle East Department at Israel's Foreign Office, to his father, Eliahu, the ambassador in Rome.[21] The younger Sasson felt that apart from uncertainties following the Iraqi revolution, Turkey had also given up hope of making Egypt's Nasser—the leader of the anti-Western camp in the Middle East— change his ways. Nasser's decision to unite his country with Syria, thus creating the United Arab Republic, gave every indication to the contrary. Turkey was now willing to resume its close ties with Israel. The sentiment was shared by Israel's minister in Washington, D.C., Ya'acov Herzog, who wrote to Moshe Sasson on 7 June 1959: "For the first time since the Nasserite threat emerged, we have found common language, and more importantly willingness for joint action, with a Middle Eastern player, and the most important one at that, from a Western point of view. Cooperation with the Turkish ambassador in Washington is a very important asset for us, giving our campaign against Nasserism a regional significance."[22]

And indeed, Turkey and Israel now had common ground in their mutual desire to develop a pro-Western alliance in the Middle East. Contacts were intensifying, culminating in a visit by Prime Minister Ben-Gurion to Ankara in late August 1958. The visit was kept secret for twenty years; only in 1977 was the visit revealed, but the records are still secret.[23] It is obvious, however, that in Ben-Gurion's talks with Menderes the foundations were laid for "peripheral" cooperation among Turkey, Iran, Ethiopia, and Israel—the non-Arab nations surrounding the Middle East (Pakistan, eager for Arab support in its ongoing conflict with India, chose not to become involved). The informal pact had its ups and downs, but it endured until the 1970s, when Emperor Haile Selassie and then the shah were deposed.

The 1950s thus ended on a bright note as far as Israel's relations with Turkey were concerned. Agreements were in place not only for strategic cooperation but also for cooperation in economic areas, agriculture, and technology. Israeli exports to Turkey increased. The two nations also agreed to collaborate in the diplomatic field, including at the United Nations. It seems they even agreed on joint military action if Aden, at the mouth of the Red Sea, were to fall into Egypt's hands.[24]

A Turn for the Worse

Since the mid-1960s, as we saw, Turkey's foreign policy had been inching closer to the Arab world and away from the West, mainly because of the situation in Cyprus. When the Six Day War broke out, Turkey was given an opportunity to demonstrate its new position. During the war and in its aftermath, Turkey exemplified complete solidarity with the Arab nations. This trend became even more pronounced after the Yom Kippur War, culminating in Turkey voting in favor of a UN General Assembly resolution, dated 10 November 1975, that defined Zionism as a racist movement. For Israel that resolution was the most outrageous ever adopted by the United Nations. For many years to come, Israel judged other nations according to how they voted on the resolution (which Ambassador Hayim Herzog tore to shreds at the podium in the General Assembly hall as soon as it was adopted). Turkey was thus found wanting.

As relations between Turkey and the Arab nations tightened, pressure on Turkey to sever completely its relations with Israel were mounting. At the first Islamic summit in Rabat, Morocco, convened in September 1969, Egypt and other nations called on Turkey to cut off all contacts with Israel. The call was rejected by Turkey's representative, Foreign Minister Çağlayangil. Turkey kept rebuffing Arab efforts to influence its foreign policy, even those by Arab oil producers after the

1973 energy crisis. In 1974 and 1975, as noted earlier, heavy pressure was exerted by Iraq, Saudi Arabia, and Libya, but Turkey still felt it could keep its oil needs and its foreign policy separate. For instance, Foreign Minister Turan Güneş said in an interview that Turkey saw no connection between the energy crisis and the Arab-Israeli conflict, even though it seemed as if the crisis had been an offshoot of the 1973 war. He stated emphatically that Turkey would conduct its policy vis-à-vis Israel in the light of international law and UN resolutions rather than fluctuating economic and political circumstances.[25] Even nominally, this statement was unusually bold, bearing in mind that a few weeks earlier both the European Common Market nations and Japan had moved closer to the Arab position because of pressure exerted by the oil producers.

The Arabs were relentless. They kept up their pressure at the Lahore Islamic summit and even more so during Erbakan's visit to Saudi Arabia in April 1974, as discussed earlier. On that occasion the Saudis explained to Erbakan (whose personal views were in line with theirs) that cutting off Turkey's contacts with Israel would make the various Arab nations and the Islamic Conference very happy and would immediately improve both Turkey's economic situation and its position in the Middle East.[26] Back home, Erbakan recommended that relations with Israel be severed, but the Turkish government rejected his views. Official Turkey still insisted that it was more important to maintain continuity and independence in its foreign policies, for the sake of consistency if nothing else.[27] Besides, argued the Turks, the Arabs were negotiating with Israel at the time and had declared their willingness to make peace with that country provided it withdrew from the Occupied Territories and recognized the national rights of the Palestinian people. Turkey, it was felt, should not be required to be "holier than thou."

Substantially, however, Ankara felt constrained to keep eroding its relationship with Israel. There was no change in diplomatic relations, and the Turkish chargé d'affaires in Tel Aviv was still a veteran diplomat with the rank of ambassador, at least until February 1981. But Ankara saw to it that commercial and cultural ties became meaningless and even described Turkey's relations with Israel as "empty" or "frozen."[28] Any news about contacts with Israel raised storms of protest in the political system, as Arab pressure was maintained. In the late 1970s, it should be recalled, Turkey was struggling to obtain oil from Arab oil producers to the exclusion of any other foreign policy consideration. Amazingly, Turkey did not totally submit to those Arab producers. But in late July 1980, Israel made a move that complicated Turkey's situation.

The Jerusalem Law

The Knesset, Israel's parliament, took up a draft Basic Law declaring Jerusalem the united, indivisible, and eternal capital of Israel. The Arab-Muslim world was outraged. Turkish prime minister Demirel found it necessary, even before the bill became law, to issue a statement strongly denouncing the move and calling for "political action" to thwart it. Other Turkish politicians joined in, led—as could be expected—by the head of the Salvation Party, Necmettin Erbakan. He dubbed the law "the crime of the century"' and demanded to "uproot Israel from the Middle East."[29] His party immediately tabled a motion in parliament to sever diplomatic relations with Israel and to express no confidence in Foreign Minister Hayrettin Erkman because of his "pro-Western and pro-Israeli" policies. Erbakan was willing to forgo the no-confidence motion if Turkey cut off relations with Israel. In the press, Turkish commentators were trying to outdo each other in their expressions of rage: Israel was using "Hitleristic methods," Israel was acting "like a raging bull," and Turkey should have "no relations with [it] whatsoever."[30]

On 29 July 1980 Demirel met with the ambassadors of the Muslim nations in Ankara at their request. They demanded that Turkey join the whole world's "strong and radical" reaction, as they put it, and significantly downgrade its relations with Israel. Demirel described the Israeli move as "unwise, wrong, and contrary to international law and fundamental justice," adding that "as a holy city, Jerusalem must not be in Israel's hands. . . . The whole world must object to the Israeli decision and act against it. . . . The Turkish government believes it is an irresponsible decision and a threat to peace."[31]

When the bill became a Basic Law, Ankara immediately recalled its chargé d'affaires from Tel Aviv for consultation. A few days later Erkman explained: "No nation can acquiesce with this change in the status of Jerusalem. The Knesset's decision on this subject need not matter, because it is not internationally recognized. A [national] law that ignores international law cannot create a new legal situation, but rather a new illegal situation." Asked whether Turkey was considering breaking off relations with Israel, Erkman replied: "We have recalled our

Chargé from Tel Aviv. We shall discuss this thoroughly and only then take the necessary steps. Severance of diplomatic relations is the final stage. We may have to go through several other steps before we get to this one."[32]

By announcing that it was reconsidering its relations with Israel, Turkey in effect invited—by design or otherwise—further pressure by the Muslim nations, particularly the oil producers. Domestic pressure increased as well. Thus Erbakan informed the Turkish public that Demirel's "pro-Israeli" policy might cause Turkey to lose its oil supplies and credit lines. He said that Iraq in particular would "almost certainly" stop oil shipments and loans to Turkey unless Ankara changed its policy toward Israel by 30 August.[33]

In late August, when Saudi finance minister Shaikh Abu Khail went to Ankara to discuss the technical aspects of the Saudi loan to Turkey, he also met with the prime minister and foreign minister, outlined the Saudi position on Jerusalem, and described Saudi expectations of Turkey. In press interviews Abu Khail attacked Israel and called on Turkey to become more involved in Muslim international affairs in view of the new threats.

Turkey soon announced the closure of its two consulates in Jerusalem, much to the satisfaction of the Arabs. A delegation of Muslim states' ambassadors again called on Demirel, this time to thank him. Libya's foreign minister announced that Turkish reaction to the Jerusalem Law had strengthened its ties with Libya. It seemed for a while that external pressures had relaxed.

Yet domestic pressures were still exerted, especially by Erbakan's Salvation Party. On 5 September 1980 it tabled a motion in parliament to remove Foreign Minister Erkman from office for various reasons, including his refusal to sever Turkey's relations with Israel. Ecevit's Republican People's Party supported the motion, Erkman's Justice Party abstained, and the motion was carried by an overwhelming majority of 230 to 2. But little happened on the political front until the 12 September coup d'état, led by General Kenan Evren, outlawed all political parties. Turkey, it seemed, was preoccupied with more urgent matters than its relations with Israel.

But even though domestic pressures ceased, with most politicians now behind bars external pressures resumed, hoping to influence the junta. Erbakan was removed from the scene for the time being, but so was Demirel, who had done his best to avoid any irreversible damage to Turkish-Israeli relations. The Arab nations stormed into the breach.

Then in late September 1980 the entire scene changed again with the outbreak of the Iran-Iraq War. I have discussed in detail its economic impact on Turkey, but the war also had a political impact, raising new uncertainties regarding oil supplies at a time when it seemed definite that Turkey would adjust its foreign policies to meet its energy needs. Consequently, on 26 November 1980 Turkey informed Israel through diplomatic channels of its decision to downgrade the level of diplomatic representation in Israel, and it asked Israel to do the same. (That same day Saudi Arabia announced that the outstanding part of its $250 million loan would soon be transferred to Turkey.) Turkey then recalled its entire diplomatic staff, except for the second secretary, from Tel Aviv and requested that the Israeli diplomatic staff, again except for the second secretary, leave the country within three months.

And so, after years of resisting Arab pressures, Turkey made a dramatic change in its formal relations with Israel in hopes that doing so would improve its economic situation and reduce its vulnerability to fluctuating oil supplies. The Turks were aware that such a move would outrage the United States, but the new foreign minister, İlter Türkmen, managed to persuade the new rulers that maintaining good relations with the Arab world was worth it.

This was a watershed in Turkish-Israeli relations, which now were nearly frozen. Additional developments in the Arab-Israeli conflict—especially Israel's decision to impose its law on the occupied Golan Heights (an act short of annexation, but not by much) in December 1981 and its invasion of Lebanon in June 1982—further complicated the situation. Even before these events, Turkey's trade union president, İbrahim Denizcier, was sharply criticized for his decision to visit Israel as a guest of the Histadrut, the Israeli Trade Union Federation, in September 1981. A Foreign Office spokesperson hastened to explain that the visit had been decided on without the government's knowledge or blessing. In 1984, when four members of parliament from the opposition visited Israel, they were criticized even within their party. Minister of State Yılmaz declared that they were hurting Turkey's political goals by going to Israel. On 23 March 1984 Prime Minister Özal announced that Turkey's relations with Israel were at their lowest point ever. Even when Israel offered to help Turkey in its struggle against terrorism, it failed to impress the Turks.

A Turn for the Better

Only in 1985, when Israel ended its siege of Beirut and began with-drawing its forces from Lebanese territory and Turkey was reducing its dependence on the oil-producing countries, did bilateral relations begin to improve. Several Turkish dignitaries visited Israel with impunity in 1986, including former foreign minister Erkman—who had been removed from office just before the coup, it will be recalled, for his support of Israel, among other sins. In September 1986 Turkey, while still keeping the appearance of low-level diplomatic representation, sent a senior diplomat, Ekrem Esat Güvendiren, to Tel Aviv; Yehuda Milo, the Israeli representative in Ankara, also had a higher personal rank than the perfunctory second secretary level retained by both diplomats. Asked about Güvendiren's appointment, the Turkish foreign minister explained in a press briefing that it had to do with "new developments in the search for peace in the Middle East."

In September 1987 the foreign ministers of the two countries met at the UN General Assembly for the first time after a long break. But the road ahead was not yet clear of obstacles. In late 1987 the intifada, a popular uprising of the Palestinians in Israeli-occupied territory, broke out. Ankara made several statements denouncing Israeli oppression and supporting the Palestinians' right of self-determination. In November 1988, when the PLO declared the establishment of a Palestinian state, Turkey immediately recognized it—the first nation to do so among those states that had diplomatic relations with Israel. The Israelis were furious and formally protested to the Turkish representative in Israel. Still, unlike most nations that recognized the Palestinian state, the Turks did not grant full diplomatic status to the PLO representative in Ankara. Speculations were rife about a possible combined deal—diplomatic status for the PLO alongside an exchange of ambassadors with Israel. In May 1990, when Greece and Israel exchanged ambassadors for the first time in history, it became clear that the Turks could not lag far behind.

Trade was also improving. During most of the 1980s it had been a modest $30–$50 million a year, but in 1989 it reached $100 million, two-thirds of which constituted Israeli exports to Turkey.[34] Several official visits took place during the year, and tourism also increased. In 1989 more than 100,000 Israelis took their vacations in Turkey, yielding about $100 million in revenue. Thus Israeli tourism was becoming a factor Turkey had to take into account.[35] Turkey's national airline, THY, resumed service to Tel Aviv, and academic contacts also expanded. Still, Turkey's relations with Israel had not reached the depth and warmth that had characterized them in the early 1950s and again in the early 1960s.

And then Süleyman Demirel became prime minister again. He has always had a warm spot in his heart for Israel, as he told me:

> For 400 years we'd ruled Palestine, offered shelter to Jewish refugees on several occasions, and treated well all Jewish communities in Turkey, even in the most remote towns. The Jews here are loyal citizens, contributing much, and we are proud of them. All this has to do with the fact that we are cautious about our relationships with you. From time to time we've reacted, we've frozen things, but on the whole we've been careful not to hurt Israel, and we shall continue to be careful.[36]

It was little surprise, then, that Demirel upgraded diplomatic relations to the full ambassadorial level in January 1992. In the 1980s Israeli diplomats in Turkey had been subject to various restrictions and limitations, but they were removed overnight.

That move, soon followed by significant progress in the Middle East peace process, brought about an unprecedented improvement in Turkey's relationship with Israel. The Turkish military became deeply involved. For years it had been leading the secular struggle against growing Islamization and was also frustrated with the West's refusal to provide it with arms and technical assistance because of Cyprus. Consequently, the military was eager to tighten its links with the Israelis. Economic relations were also thriving. Israel became one of Turkey's closest allies, if not its closest.

Between June 1992 and November 1994 a series of unprecedented high-level visits took place. Israeli president Ezer Weizman and foreign minister Shimon Peres went to Turkey, and Turkish prime minister Tansu Çiller visited Israel. Turkey firmly asserted its "right" to have close ties with Israel, rebuffing all Arab and Muslim protests.

In March 1996 Turkey's president Demirel made an official visit to Israel. Taking place a few months after the assassination of Prime Minister Yitzhak Rabin, the visit was intended to display Turkey's support for the peace process in the Middle East and to encourage Peres, who was now prime minister, to pursue the policies of his late predecessor. During the visit Israel and Turkey signed a series of economic agreements calling for the removal of all customs barriers on the trade between them by 2000 and more cooperation in agriculture, industry, telecommunications, and medicine, with an emphasis on sharing advanced technologies.

Turkey's relations with Israel acquired yet another dimension, one that had been dormant since the demise of the 1958 strategic cooperation agreement: military ties became closer. For instance, the Turkish air force made an agreement with Israel's aeronautical industries to renovate its F-4 Phantom jet fighters. Similar agreements opened Turkish air and naval bases to Israeli military aircraft and vessels.

According to various reports in the Israeli press, military cooperation between Turkey and Israel became intensive. Quoting Turkish deputy chief of staff General Çevik Bir, newspapers spoke of a joint strategic study group being set up, of intelligence information being exchanged, of intensified links forming between the two countries' air forces and navies, of Israeli military personnel assisting the Turkish military in various organization and training activities, and of numerous high-ranking officers making mutual visits—including as observers watching various exercises.

Thus in January 1998 the highly publicized Mermaid naval exercise took place in which Turkey's and Israel's navies, as well as the U.S. Sixth Fleet, took part. Jordan sent an observer. Described as only a "search and rescue" exercise, it nevertheless drew many protests from Muslim nations.

To the extent that they officially referred to their military cooperation at all, Turkey and Israel always described it as purely bilateral, not aimed against any particular country or group of countries. Nonetheless, as media reports multiplied and were rarely denied, the situation became increasingly awkward. Particularly worrisome for Israel were reports that Israel had helped Turkey acquire intelligence information on the whereabouts of Kurdish guerrilla leader Abdullah Öcalan, leading to his capture by Turkish security personnel in Kenya. The reports, mostly in the Turkish press, caused much anger among the Kurds' supporters, and Kurdish threats of retaliation were made. In this case, the information was vehemently denied by Israeli spokespersons.

But other aspects of military cooperation were crystal clear. The F-4 enhancement deal was soon followed by an F-5 deal. Other reports spoke of the supply of Israeli-made Python-4 air-to-air missiles and Popeye air-to-surface ones, joint development of a cruise missile named Delilah against antiaircraft and maritime surface-to-surface missiles, supply of Israeli night sights for helicopters, possible supply of Israeli Arrow antimissile missiles, and more.

A report in the Israeli press uncovered yet another development: collaboration between the U.S. General Dynamics Corporation and the Israeli Military Industries to upgrade 1,000 of Turkey's M-60 tanks. Israel had hoped to sell Turkey its own Chariot tanks, but nevertheless it left Israel a partner in a gigantic deal, amounting to $7 billion.

All this severely strained Turkey's relations with the Arab world. Syria felt itself in the grip of a Turkish-Israeli pincer; Iran and Iraq clamored against this alliance with Zionism; and even more moderate nations, some of which had diplomatic relations with Israel (such as Egypt), were severely critical of Turkey. Thus Iran suddenly "discovered" a spy ring in Turkish consulates in its territory and immediately linked it to joint Turkish-Israeli military exercises. The Iranian press suggested creating an all-Arab front to combat Turkey's cooperation with Israel. Syria described the situation as a "dangerous game." Lebanon's Shaikh Hassan Nasrallah, the leader of the Hizballah movement, warned the Turks of Israel's ulterior motives: it was seeking to control Turkey's water resources, finances, and decisionmaking processes. A senior Egyptian official, Usama al-Baz, worried that the Turkish-Israeli alliance might make Israel a military power capable of imposing its will on the entire Middle East through the force of arms.

The Turks, although endeavoring to reassure the Egyptians and others that their alliance with Israel was not aimed against anyone, nonetheless made no effort to deny its existence or downplay its scope. In the face of Arab and Muslim outrage, the Turkish military confirmed that Israeli pilots had already begun exercises in Turkish air space and that naval vessels were in Turkish harbors, per the agreement between the two countries. Foreign Minister Emre Gönensay said in 1996 that Turkey was under no obligation to report to anyone regarding agreements it was making to enhance its defense, although he denied the existence of a strategic agreement with Israel. A short while later Prime Minister Yılmaz announced that Turkey had no intention of yielding to Arab pressure on this issue.

The Turkish ambassador in Israel, Barlas Özener, told the *Jerusalem Post* (1 May 1996): "We do not want to deal with Israel behind closed doors any more. . . . We object to people whose governments made peace with Israel describing it as 'the Arab world's greatest enemy' or inferring that Turkey is engaged in nefarious 'plots' with Israel."

This rapprochement was highlighted when Israeli prime minister Ehud Barak made an official visit to Turkey in October 1999. The main purpose of the visit was to inaugurate a village built with Israel's help for victims of the severe earthquake that had hit Turkey a few months earlier (discussed later), but he also held high-level talks with Turkey's leaders, concluding a number of economic and strategic agreements and discussing others—including the possibility of exporting water from Turkey to Israel. Forty-one years after Ben-Gurion's secret visit to Ankara, an Israeli prime minister visited again, this time as an official, honored guest. Fifteen years after Turkey's prime minister announced that relations had hit rock bottom, Barak was welcomed with the respect the leader of a particularly friendly nation deserved. A complete transformation had taken place.

Notes

1. *Hürriyet*, 10 April 1986.
2. Turkish State Institute of Statistics, *Annual Foreign Trade Statistics*, various years.
3. See, for instance, *Günaydin* and *Tercüman*, 23 November 1989.
4. Former Turkish ambassador to the United Nations Coşkun Kırca made such a call in *Milliyet*, 27 December 1989.
5. See *al-Hayyat* (London), 7 February 1990.
6. Quoted in all Turkish newspapers, 17 September 1998.
7. *Yevi Yuzyil*, 2 October 1998.
8. The Öcalan saga continues in Chapter 11.
9. *Milliyet*, 24 April 2000.
10. The Turkish parliamentary opposition objected unanimously to the visit. See statement by its leader, Bülent Ecevit, quoted in *Cumhuriyet*, 3 December 1977.
11. *Cumhuriyet*, 2 December 1977.
12. *Milliyet*, 3 and 8 December 1977.
13. See, for instance, *Pulse*, 9 December 1977.
14. A. Nahamani, *Israel, Turkey, and Greece* (London: Frank Cass, 1987), p. 56.
15. Minister in Ankara to the Foreign Office, 25 September 1953, Israel State Archives, Jerusalem (henceforward ISA).
16. Cable, Ankara to Jerusalem, 23 September 1957, ISA.
17. G. E. Gruen, *Turkey, Israel, and the Palestinian Question, 1948–1960* (Ph.D. dissertation, Columbia University, 1970), pp. 342–344.
18. Cable from Rome to Jerusalem, 25 September 1957, ISA.
19. Internal memo, Israeli Foreign Office, November 1957, ISA.

20. Israeli Embassy in Ankara to the Ministry in Jeruslam, 23 September 1957, ISA.

21. Ministry of Foreign Affairs in Jerusalem to the Israeli Embassy in Rome, 25 May 1959, ISA.

22. Israeli Embassy in Washington to Israeli Embassy in Rome, 7 June 1959, ISA.

23. M. Bar-Zohar, *Ben Gurion* (Hebrew) (Tel Aviv: Am Oved, 1977), Vol. 3, pp. 1329–1330.

24. Nahamani, *Israel, Turkey, and Greece,* p. 75.

25. *Cumhuriyet,* 17 February 1974.

26. See *Pulse,* 24 April 1974.

27. Ayhan Kemal, "Turkey's Relations with the Arab World," *Foreign Policy* (Ankara) 4, no. 4 (1974), p. 102.

28. See ibid.; *Pulse,* 21 July 1975.

29. *Cumhuriyet,* 29 July 1980.

30. *Hürriyet,* 27 and 26 July, *Cumhuriyet,* 27 July 1980, respectively.

31. *Pulse,* 30 July 1980.

32. *Tercüman,* 4 July 1980.

33. Statement made at a press conference upon his return from Baghdad.

34. Israeli Government Annual Report (Jerusalem: Prime Minister's Office, 1990).

35. This trend has continued. At present, Turkey is still perhaps the destination of choice for Israeli vacationers.

36. Interview with the author, Ankara, 20 May 1985.

PART THREE

TURKEY TOWARD THE TWENTY-FIRST CENTURY

10

THE ISLAMIC CHALLENGE

The 24 December 1995 general elections turned Turkey's political system upside down. In the most significant political transformation in Turkey since the days of Mustafa Kemal Atatürk, Islam returned to center stage in Turkey's national life. The Refah Partisi (RP, Welfare Party) won more than 21 percent of the total vote, more than each of the other four political parties that managed to get their representatives elected to parliament.[1] That percentage translated into 158 seats in the 550-member house. Thus for the first time in seventy-two years of secular democracy, an Islamic party won the general elections.

Two right-of-center parties followed closely, with about 19 percent each. The Doğru Yol Partisi (DYP, True Path Party), led by Tansu Çiller, won 135 seats, and Mesut Yılmaz's Anavatan Partisi (ANAP, Motherland Party) won 132 seats. Both came in ahead of the left-wing parties: Bülent Ecevit's Demokratik Sol Parti (DSP, Democratic Left Party) won 76 parliamentary seats, and Deniz Baykal and the Cumhuriyet Halk Partisi (CHP, Republican People's Party) finished last, with 49 seats. Several other political parties, although failing to reach the steep 10 percent threshold, nevertheless scored remarkably well: the extreme right-wing pan-Turanic Milliyetci Hareket Partisi (MHP, Nationalist Movement Party), led by Alparslan Türkeş, won 8.2 percent of the vote, and the Kurdish Halkin Demokrasi Partici (Hadep, People's Democracy Party) managed 4 percent of the total vote despite all obstacles.[2]

219

As soon as the results became known, the RP launched an intensive (and fairly successful) effort to improve its image in Turkey and abroad. It downplayed its familiar positions—especially its calls to strengthen Turkey's ties with the Muslim world, to renegotiate the customs agreements between Turkey and the European Union (EU), and to reexamine Turkey's North Atlantic Treaty Organization (NATO) membership. The party's leader, Necmettin Erbakan, was trying to persuade public opinion in Turkey and the West that there was no reason to fear either the RP or a government led by the party. In a complete about-face, bearing in mind his decades of fierce anti-Western rhetoric, Erbakan now emphasized that under his leadership Turkey would maintain its close relationship with the Western world and keep its market economy, as well as its agreements with the EU.

At the same time, the RP strove to maintain its reputation as an uncompromising enemy of corruption in public affairs, because in the 1995 elections it had rallied around itself protest votes from both inner cities and remote rural areas, in addition to its regular pro-Islamic voters. Tom Friedman regarded this as an "antiglobalization" reaction: "Those who understood the new game [globalization] flourished. Those who didn't got left behind . . . creating a huge urban underclass" that gave the RP its victory.[3]

Shortly after the December 1995 elections the secular political parties, supported by the military and the business community, made a strenuous effort to keep Erbakan out of the coalition, despite his displays of newly found moderation. Both right-of-center parties, the DYP and ANAP, had secular, pro-Western platforms. And since both left-wing parties had also raised the banner of secularity, it meant that nearly 80 percent of voters supported Turkey's continued secularity following the Kemalist principles.

It was only natural, therefore, for both moderate right-wing parties to unite immediately after the elections into a single political bloc that would have led a secular coalition, but that was not to be. The intense personal rivalry between Yılmaz and Çiller made it impossible for them to join forces, much less agree on who would lead a coalition government. Although both leaders yielded to the enormous pressure exerted by the business community and created a conservative coalition under Yılmaz (on 13 March 1996), it lasted only three months before collapsing when corruption allegations were leveled at Çiller.

The RP made much of the allegations. On 9 May 1996 parliament overwhelmingly decided to set up a parliamentary commission of inquiry into possible acts of personal corruption by Çiller. Most ANAP representatives supported the resolution, even though the party was Çiller's coalition partner. That was the final straw. On 3 June 1996 the Yılmaz-Çiller coalition lost a no-confidence vote in parliament. Three days later President Süleyman Demirel summoned Erbakan and asked him to create a new government.

Erbakan's Government

On 28 June 1996 the RP announced that it had managed to set up a coalition with Çiller's DYP. The coalition was based on a rotation of the office of prime minister: Erbakan would hold the position for two years, then Çiller would take over for another two years. For many Turks, brought up on Atatürk's legacy, this was a nightmare come true. Erbakan—a minor politician in the 1970s and a political nonentity for most of the 1980s—came back to the helm of the nation in the mid-1990s. The politician who in the 1970s had stood for uncompromising fundamentalism and persisted in taking anticonstitutional positions would lead Turkey into the twenty-first century.

Erbakan's party had always been concerned with economic and social affairs, and he kept for its leaders the finance, justice, and culture portfolios. Çiller became deputy prime minister and won for her party the foreign affairs and defense portfolios, as well as education after tough negotiations.

Ever since the establishment of the Turkish Republic in 1923, secularity had been a cornerstone of its modern national identity. In 1937 it became an article of the constitution; any attempt to change Turkey's secular regime thus became a criminal offense. And now, in June 1996, Turkey found itself led by a person whose political philosophy directly challenged that principle and indeed the constitution itself.

In terms of foreign policy, things also became confused. Relations with the Arab nations and the Muslim world as a whole had been kept on the back burner since the mid-1980s, but now the champion of Turkey's rapprochement with the Muslim world held the reins of power. And conversely, at a time when Turkey's relations with Israel were unprecedently close, Israel's sworn enemy became prime minister. During the 1970s and 1980s Erbakan's anti-Israel campaigns, which included anti-Semitic elements, had repeatedly harmed Turkey's relations with Israel. Now Erbakan found Israel to be Turkey's closest ally in the region, if not in the entire world.

Erbakan's long-standing and familiar vision for Turkey—to establish a new cultural order—was quickly redefined to make it more suitable for the administration of a modern country. The need to do so was especially urgent because of the ever-present threat of yet another military coup d'état. Erbakan thus declared his adherence to the principle of secularity and his support of continued ties with the West (although he called for better relations with Balkan and Muslim nations), announced that he stood for a market economy, and even paid tribute to Atatürk. The new prime minister defined his government's guidelines as "democratic, secular, and egalitarian" in the spirit of Atatürk's principles. Such moderate pronouncements, along with the DYP's participation in government and the military's growing involvement in civilian life, put to rest for a time fears in Turkey and abroad that Islamic values might be coercively imposed on the entire nation.

In Erbakan's first address to parliament as prime minister on 3 July 1996, the words *moderation* and *continuity* were repeated often. The plans he introduced to combat terror, inflation, unemployment, and poverty were noncontroversial. His aim, he said, was to restore peace, confidence, and hope for all Turks. On the Kurdish issue he expressed support for a military solution, in line with Turkey's secular leaders, even though some of his best friends in the Muslim world—the Syrians first and foremost—actively supported the Parti-ye Kerkaran-î Kürdistan (PKK, Kurdistan Workers' Party), the Kurdish underground movement.

On 9 July 1996 Erbakan's government received parliament's vote of confidence, albeit with a small majority (278 to 265). Eight DYP members crossed the floor in protest, thus weakening Çiller vis-à-vis her coalition partner.

Despite all moderate statements to the contrary, the RP never gave up its hope of transforming all of Turkish society. Erbakan's government channeled its Islamic ambitions mainly into the education and cultural spheres, antagonizing the military as well as the secular civilian public in the process. Leading this effort were the RP's three most prominent ministers: Abdüllatif Şener in the Treasury, Minister of Justice Şevket Kazan, and Minister of Culture Ismail Kahraman. Under Erbakan, they were seeking to reform education, culture, religion, the legal system, and the economy. They placed their appointees on the boards of trustees of universities and museums; they financed religious schools at the expense of secular ones; they reduced compulsory state-provided education from eight to five years to provide more scope for private education; they proposed to set up mosques in places regarded as shrines of secularity, such as Istanbul's Taksim Square and Agia Sophia Museum; on official occasions they preferred to quote religious writers (including leaders of the various orders) rather than Turkey's leading literary lights, such as novelists Yaşar Kemal and Orhan Pamuk; they sought to set a ceiling on interest rates (interest is forbidden by Islam) and to establish Islamic economic institutions; and they rewarded prisoners with improved conditions for citing verses from the Quran. Also, they tried to block legislation against in-family violence, which they regarded as illegitimate interference in personal affairs.

Most outrageous in the eyes of the military was an invitation extended to the leaders of the Sufi orders to break the Ramadan fast at Prime Minister Erbakan's official residence—disregarding the fact that Turkish law still banned those orders. Erbakan also refused to have the national anthem sung in his presence by the national choir (which was coed); he always carried a cassette on which the anthem was sung by a male choir.

In early 1997 the military leaders felt the situation was foreboding. The chief of the Turkish navy, Admiral Güven Erkaya, declared that Islamization was the greatest threat to Turkish national security—greater even than that of the PKK.[4] Chief of Staff General Ismail Karadayı warned that the RP was taking Turkey "back to the middle ages." He informed Erbakan that the military was keeping an eye on his government's activities in the sphere of education and would see to it that no red lines were crossed.

Yet Islamization continued. Finally, on 24 March 1997, the National Security Council sent Erbakan a list of eighteen demands aimed at curbing the growing Islamization of Turkey. The document emphasized that the Sufi orders were still outlawed, demanded eight years of compulsory (state-provided) education and a limit on the number of students in religious schools, demanded that fundamentalists be removed from the military, called for an end to Iranian subversive activities (namely, support for the spreading of Islam), warned against private armed militias (a clear reference to Erbakan's RP-recruited bodyguards), stressed that the legal code of dress outlawed veils for women in public places, and more.[5]

Established after the 1980 coup, the National Security Council in effect has ultimate authority over security and defense issues in Turkey. Consisting of half military and half political leaders, the council is headed by the state president and includes the prime minister, as well as the ministers of defense, foreign affairs, and internal affairs. Yet the military's influence has always been dominant.

Erbakan tried to procrastinate but eventually undertook to fulfill the demands. But the military leadership soon decided he was not abiding by his promise and appealed to the State Security Court, asking that he personally and his party as a whole be barred from political activity.[6] At the same time they worked behind the scenes, along with leaders of the business community, to set up an alternative coalition.

Thus in July 1997 Erbakan's government fell. It was replaced by a coalition led by Mesut Yılmaz and the ANAP, which this time brought in as their partner the left-wing DSP under Ecevit, who became deputy prime minister. The coalition was supported "from the outside" by Baykal's CHP. Properly speaking, this was not a coup d'état. It was different from the moves made by the military in 1960, 1971, and 1980 in that one parliamentary coalition substituted for another. This time the military was more subtle—but it still achieved its aim.

Erbakan's Foreign Policy

Necmettin Erbakan's record as prime minister proves that the Islamic RP, even as a coalition leader, was unable to redirect significantly Turkey's foreign policy. He made serious efforts to move closer to such "sister nations" as Iran, Iraq, and Libya, but his work brought about no change in Turkey's global orientation. Turkey's commitment to the West remained intact, despite growing friction in its relations with the EU.

Economically, too, Erbakan was unable to implement the principles embodied in his party's platform. In virtually all respects, his government's economic policies were merely a continuation of those of the previous government. He realized that Turkey's future as an industrial nation depended on selling abroad finished products rather than raw materials and that economic survival entailed selling to the West. As much as he wanted to, he could do little about Turkey's strong economic links with the West.

In one area in particular the RP did its best to impose its beliefs and principles on Turkish foreign policy: Turkey's relations with Israel. The Muslim nations' objections to the tightening links between the two countries in the military sphere were noted earlier. They were echoed in Muslim public opinion in Turkey, led by RP leaders. Things got out of hand in May 1996: when President Demirel visited the town of Izmit shortly after his return from Israel, a pharmacist named İbrahim Gümrükçüoğlu made an attempt on his life.

Upon becoming prime minister, Erbakan announced that Turkey would respect all international agreements it had entered in the past, "unless they contradict the interests of national security."[7] His foreign minister and coalition partner, Tansu Çiller, immediately sent a reassuring message to Israel: this was not a reference to Turkey's agreements with Israel. Indeed, Erbakan soon found out that even here he was unable to conduct foreign policy as he wished in the face of the military's determination to maintain intensive relations with Israel.

Gradually, Turkey's foreign policy became divided: Erbakan handled, in his own style and according to his own line, Turkey's relations with the East and with the Muslim world, leaving the West and Israel to Çiller. In a series of visits to Muslim nations, Erbakan tried to convince the Turkish public that their nation could enjoy the best of both worlds—strong economic ties with the West and close cultural, political, and economic links with the Muslim nations. There was no contradiction here, Erbakan maintained, not even with membership in NATO while striving at the same time to create a "Muslim NATO."

Shortly after coming to office Erbakan visited Iran, where he made comprehensive agreements for gas imports and expanded bilateral trade. Iranian president 'Ali Akbar Hashemi Rafsanjani, in late 1996, made a return visit that became rather controversial: he spoke highly of Turkey's "return to Islam" and refused to abide by protocol and lay a wreath at the Atatürk monument. Despite this, more agreements were made during his visit. But the military rejected the idea of a military agreement with Iran as a counterbalance to the Israeli agreement.

The agreements the Erbakan government signed with Iran outraged the United States; during Erbakan's visit to Teheran, Washington saw fit to remind everyone that it would consider sanctions against any nation or corporation trading with Iran. Richard Perle, a former senior Pentagon official, said that Erbakan "stuck his finger in America's eye,"[8] and Tom Friedman wrote in the *New York Times* (21 August 1996) that "suddenly Turkey's government, a longtime pillar of U.S. policy at the crossroads of Asia and Europe, is becoming, under Mr.

Erbakan, an unreliable ally at best and a threat to U.S. interests at worst."

Erbakan also attempted to improve relations with Iraq. A high-level delegation went from Ankara to Baghdad in August 1996, led by the ministers of oil and industry. The Turks were willing to become involved in the UN-sponsored "oil for foodstuffs and medical supplies" deal. Western economic sanctions on Iraq were a burden to the Turkish economy, and Turkey under Erbakan was pressuring Washington to allow it to trade with Iraq outside the sanctions.

Another initiative taken by Erbakan was no less controversial—his attempt to improve relations with Libya. Few foreign leaders had visited Tripoli since UN sanctions were imposed on Libya in 1991, but Erbakan decided to go there in October 1996—both to rectify this injustice, as he saw it, and to improve economic relations. In this respect, an important item on his agenda was collecting a $365 million Libyan debt to Turkey.

The visit was a huge fiasco. Erbakan had barely arrived when Muammar Qaddafi declared, "the Kurdish people deserve independence as much as the Arabs do."[9] The Turkish public was furious, and Erbakan narrowly escaped a vote of no confidence following the visit. He did not escape a formal protest served by the U.S. ambassador against both the visit and the economic agreements concluded in its course.

The Turkish military stood in the breach during Erbakan's term of office. Thus the military "persuaded" Erbakan to conclude the F-4 upgrade deal with Israel in December 1996, worth approximately $650 million, and had him meet with Israeli foreign minister David Levy—a tense and chilly meeting that did warm up as it progressed. All in all, the military's desire to maintain close relations with Israel was a major obstacle to Erbakan's plan to overhaul Turkey's Middle Eastern policies. Eventually, as related earlier, the military brought Erbakan's reign to an abrupt end.

The Susurluk Affair

In early November 1996 Turkey was shaken by a political scandal that continues to reverberate. In a car accident near the resort town of Susurluk (100 miles southwest of Istanbul), three passengers in a speeding car were killed and one was injured. When their identities became known, Turkey was shocked; dark and evil secrets were suddenly uncovered. One of the victims was a wanted criminal named Abdullah Çatlı, another was a former Miss Turkey, and the other was a senior police officer. The injured man was a member of parliament for the Doğru Yol Partisi, Sedat Bucak, who was also the head of Turkey's largest private militia. His organization was used by the authorities to combat the PKK for an annual consideration of $1.3 million. A cache of weapons and forged passports was found at the accident site.

This unsettling combination of organized crime, the police, glamour, and politics led directly to Minister of the Interior Mehmet Ağar, who allegedly had spent the weekend prior to the accident at a seaside resort with the four victims. Despite all efforts by DYP leader Çiller to dissociate herself and her party from the affair, the minister was soon forced to resign.

The investigation led to more suspicions of collaboration between organized crime and senior politicians. Çiller, already the target of many other accusations of corruption, saw her reputation tarnished further as a consequence of this affair. All of this was grist to the mill of Prime Minister Erbakan, who accused the leaders of the previous government of using "state gangs," as he put it, in the struggle against the Kurds. "Nothing," said Erbakan, "including our fight against the PKK, can justify [the] commission of crimes."[10]

The Susurluk affair became even more complicated when it was discovered that Çatlı, a criminal wanted by both Turkish police and Interpol, held a Turkish diplomatic passport as well as six identity cards—each under a different name. Several senior police officers, including the chief of the Istanbul police, were asked to resign. Still, allegations of criminal involvement in Turkey's ongoing struggle against the PKK have not been fully put to rest.

A Left-Right Government

The Erbakan-Çiller government was replaced in July 1997 by a coalition headed by Mesut Yılmaz and the ANAP, with Bülent Ecevit and the DSP as their major partners. The most surprising aspect of this unlikely coalition between the right-wing ANAP and the left-wing DSP was the harmony among its leaders. Despite ideological differences, Yılmaz and Ecevit were able to work together and stabilize Turkish politics to an extent.

The military, not content with removing Erbakan from power through the political process, petitioned to bar him from politics for good and to outlaw his RP as a threat to the nation's secular regime. The appropriate venue for such a petition in Turkey, as noted, is the State Security Court. And indeed, in January 1998 the court decreed that Erbakan and four other senior RP leaders be barred from politics for five years and that the largest party in parliament be dissolved.

Such injunctions were nothing new to Erbakan, then seventy-three years old. He regrouped immediately. A newly formed organization, Fazilet Partisi (FP, Virtue Party), headed by Recai Kutan, soon appeared. Unsurprisingly, it included many RP members of parliament, who thus remained highly influential as the largest faction in that body. It was essentially the same party with Erbakan as the mover and shaker, albeit behind the scenes.

The military, unfazed by this change of guise, continued its struggle against Islamization under the new government. Prime Minister Yılmaz called on the officers to mind their business and let the government combat religious fanaticism on its own, but the military remained adamant. Yılmaz well knew who had put him in power, and in March 1998 he reached an agreement with the military, under the auspices of the National Security Council, to enforce "uncompromisingly" existing laws against religious excesses. That meant first and foremost the ban on women's veils in state educational institutions, an issue that had become a symbol of the entire struggle against Islamization.

The military had been able to use the National Security Council as an instrument in the struggle because it had the wholehearted support of President Süleyman Demirel, who had the deciding vote. In an otherwise deadlocked council, composed of politicians and military officers in equal numbers, President Demirel has been a key figure in the defense of secularity.

On 23 April 1998 the State Security Court handed down a ten-month jail sentence to Recep Tayyip Erdoğan, the mayor of Istanbul and one of the leading Islamic activists in the country (some regard him as Erbakan's heir apparent). His crime was giving a speech to an audience of activists, whom he compared to an army: "The mosques are our bases, their domes are our helmets, their minarets are our lances, the faithful are our soldiers."[11] The military could not allow such rhetoric in Turkey.

One offshoot of the military-led struggle against Islamization has been a growing tension between Sunni and 'Alawi Muslims. The

'Alawis have always been regarded by fundamentalists as supporters of secularity. Members of this minority group (around 20 percent of Turkey's population) do not often frequent the mosques, and many do not fast in the month of Ramadan. Even Erbakan, when he attempted to negotiate with their leaders to regularize relations with Turkey's religious authorities, was thwarted by the fundamentalists. In 1998 there were more than thirty 'Alawi representatives in the Turkish parliament, most belonging to the CHP—once headed by Atatürk himself.

In May 1998 thirty-three Sunnis were sentenced to death for setting fire to a hotel where 'Alawis had gathered to celebrate the memory of one of their leaders. Thirty-seven died in the fire, described by the court as "the most abominable event in the history of Islam in Turkey."[12] The arsonists were unrepentant. This brought tensions between Sunnis and 'Alawis to a new level.

The Virtue Party

In an almost exact recapitulation of the emergence in 1972 of the Milli Selamet Partisi (National Salvation Party) and in 1983 of the Refah Partisi, the old-new Fazilet Partisi took shape as soon as its predecessor had been disbanded. Once again its new leader, Recai Kutan, tried to create for his party a moderate image as a supporter of virtue rather than an enemy of democracy. He wished that secularity would not become "an instrument of oppression against the freedom of conscience and religion. . . . We do not object to secularity," said Kutan, "we only object to secular obsessions."[13]

Emphasizing democracy, supporting a market economy, objecting to government involvement in economic affairs, supporting the effort to make Turkey a member of the European Union, and placing the FP as "neither extreme left wing nor extreme right wing but rather center," the party's platform sought to wipe out its fundamentalist public image. Furthermore, such statements sounded more believable coming from the relatively unknown Kutan than they had when uttered by Erbakan, whose views were known to all Turks. The main differences between the FP and other parties have been in the spheres of culture and education, although the FP also demanded—unsurprisingly—a "complete reorganization" (namely, dissolution) of the National Security Council "in the spirit of the values of democracy as practiced in the West."[14]

The elderly and comparatively moderate Kutan has not been as popular as Erbakan among his party's constituents. Many would have preferred the party's rising star Erdoğan, despite (or maybe because of) his jail sentence. Erbakan, however, chose a less charismatic and more obedient person to lead his party while he is unable to do so himself, probably in expectation that Kutan will step down when Erbakan's ban runs its course in January 2003. Also, the military has acquiesced with Kutan's leadership, but it is doubtful whether it would have tolerated Erdoğan.

In view of Erbakan's swift recovery from his political-legal setback in early 1998 and the effective organization of his party under its new guise as the FP, some secular politicians and even military men began to realize that Islam in Turkey is not easy to subdue. They began to contemplate the development of a "modern Turkish Islam" as a substitute for the kind of Islamic fundamentalism practiced in Iran, for instance ("the abomination imported from Iran," as Yılmaz called it). Thus military leaders began talking about "our Islam" and denouncing those who tried to besmirch it. One general, Air Force commander Erhan Kılıç, instituted the reading of verses from the Quran to his troops in Turkish rather than Arabic. That is common practice in the Bektaşi Order, which is regarded as relatively liberal and has much support among army veterans; more traditional elements frown upon it, if not regard it as plain heresy. It is small wonder, then, that the book *Islam in the Turkish Language* by Kemal Kutay recently became a bestseller.

It seems that in the early twenty-first century, Islam in Turkey is at a crossroads. Most Turks have always regarded themselves as Muslims, and the constitution has never interfered with individual beliefs. When former president Turgut Özal described his people as holding a computer in one hand and the Quran in the other, he was not far off the mark. According to official figures, about 75 percent of Turkey's adult population observes the Ramadan fast, and around 50 percent of all adult males go to one of the country's 80,000 mosques each Friday.[15]

Islamic clergy are public officials, drawing their salaries from the state treasury. As such, they are bound by the Atatürk legacy, which calls for the separation of state and religion. Political Islam, namely the Fazilet Partisi, has been constrained by military pressures and court rulings and has endeavored to portray itself as moderate and conciliatory to both domestic and international publics. Hard-core fundamentalist Islam now relies on outlawed underground religious orders, manned by volunteers and with an informal education system. Some of the orders own widely circulated newspapers, and their influence is not insignificant.

The FP's decline in the 1999 elections (discussed later) did not end the tensions between it and the military or remove the Islamization problem from the national agenda. When a newly elected member of parliament for the FP, Merve Kavakçı, wore a veil to the parliament's inauguration ceremony in May 1999, the issue flared up again. Even the leaders of her own party disowned her, and she was not sworn in. Yet this public display of denial by the FP seems to have been to no avail. At the time of writing, efforts are under way to outlaw and disband the Virtue Party once again.

The web of interrelationships among official Islam, political Islam, and underground Islam is complex, and demarcation lines are nowhere clearly drawn. The secular sector views all of these factions with varying degrees of suspicion, and the military is ever watchful for any breach.

Notes

1. Recall that under the constitution, a party has to win at least 10 percent of the total vote to have parliamentary representation.

2. Bruce Maddy-Weitzman, ed., *Middle East Contemporary Survey, 1996,* Vol. 20 (Boulder: Westview, 1998).

3. *New York Times,* 17 July 1996.

4. Aryeh Shmuelevitz, *Turkey's Experiment in Islamic Government 1996–1997: Data and Analysis* (Tel Aviv: Moshe Dayan Center, Tel Aviv University, May 1999), p. 24.

5. Ibid., pp. 26–27.

6. The State Security Court is part of the Turkish legal system. Its purview includes all matters relating to the security of the Turkish state, as well as perceived threats to the constitution. It is not, however, the highest court in the land: the Supreme Court can overturn its decisions on appeal.

7. His speech in parliament, 3 July 1996, quoted in Maddy-Weitzman, p. 707.

8. Perle is quoted in Tom Friedman, *New York Times,* 21 August 1996.

9. In a press briefing to Turkish journalists, *Ha'aretz,* 6 October 1996; TRT TV, 5–7 October 1996.

10. *Jerusalem Post,* 4 December 1996.

11. Quoted in *Ha'aretz* (Hebrew), 24 April 1998.

12. *Economist,* 25 April 1998.

13. *Turkish Daily News,* 18 December 1998.

14. Ibid.

15. Interview by the author with the imam of Istanbul in Israel, 1999.

11

GROUNDSWELL, LANDSLIDE, AND EARTHQUAKE

Throughout the 1990s Turkey persisted in describing the Kurdish issue as a military problem rather than a political struggle for the recognition of Kurdish rights. As Turkish leaders used to say, "There is no Kurdish problem, there is a terror problem." All five political parties represented in the parliament elected in 1995 favored a military solution to the problem, and thus the government was able to ignore the increasingly moderate tone in which Kurdish demands have been put forward by Parti-ye Kerkaran-î Kürdistan (PKK, Kurdistan Workers' Party) leader Abdullah Öcalan, including repeated declarations of a unilateral cease-fire.

The Kurdish Problem Re-erupts

Following the Gulf War, a window of opportunity opened briefly in early 1991 for the creation of an autonomous Kurdish entity in northern Iraq. Turkish reaction, however, was hostile when the idea was put forward. The ever-present fear that the establishment of any Kurdish entity would inflame Kurdish aspirations within Turkey once again prevailed. As Tansu Çiller put it, "We shall not allow an independent Kurdish enclave in northern Iraq. Iraq's sovereignty and territorial integrity are important to Turkey."[1]

Yet Western public opinion relentlessly harassed Turkey over the

Kurdish issue. Particularly painful were the diatribes poured out by Yaşar Kemal, a distinguished novelist living in Germany. In an article in *Der Spiegel* in January 1995, he called Turkey's struggle against the Kurds "the meanest, most abominable war imaginable." He selected for special denunciation the civil defense militias in eastern Turkish villages (numbering 50,000 men, according to him), which had operated alongside the Turkish military, and the policy of burning forests to drive out PKK guerrillas.

In late 1996 Ankara hosted talks between the two rival factions of Iraqi Kurds, Jalal Talabani's Kurdistan Patriotic Union and Mas'ud Barazani's Kurdistan Democratic Party. Both were making determined efforts to improve their positions in Kurdistan—Barazani by moving closer to Turkey and Iraq, Talabani by moving closer to Iran and Syria.[2] Shortly afterward the parties met again in the United States; Assistant Secretary of State Robert Pelletreau, who masterminded this round of talks, neglected to invite Turkish representatives. Discussions resulted in a series of political agreements between the parties, and it seemed as if Kurdish autonomy was a more realistic prospect. But Ankara hastened to announce that any deals made in Washington in Turkey's absence were unacceptable, and the talks collapsed. The West increased its pressure on Turkey to desist from violating the human rights of the Kurdish people. One particularly damaging form this pressure took was the refusal to provide Turkey with modern armaments promised after the Gulf War, as well as the denial of economic assistance.

Following increased anti-Turkish activities by the PKK in Iraqi territory, in September 1996 Turkey declared a "security zone" of its own in northern Iraq in which it saw itself free to conduct military operations. They would desist, said the Turks, only if Saddam Hussein proved capable of checking Kurdish guerrilla operations against Turkish territory.

At the same time, tensions were mounting between Turkey and Syria over the PKK. Top-ranking officials such as Süleyman Demirel and Melut Yılmaz publicly accused Syria of aiding Kurdish terrorists and of harboring Öcalan. In fact, the Turks had known this all along but felt constrained to accept Syrian denials at face value. But no longer: now they even threatened Syria's water supply unless it withdrew its support of the PKK. In the spring of 1996 a series of bomb attacks took place in Damascus, one of them near Öcalan's home. Some observers speculated that the attacks were engineered by Turkey in a kind of quid pro quo.

The Kurdish struggle has come a long way since 1973, when Öcalan established a Marxist-Leninist student group at the University of Ankara that later called itself the PKK. Since 1984 its guerrilla operations have caused the deaths of more than 30,000 soldiers and civilians, creating an unbridgeable abyss between Turkey and the Kurdish guerrillas. When Öcalan began his efforts to shift the struggle to political channels in 1993, it was too late. A series of declarations of unilateral cease-fire—in 1993, 1995, and 1998—was met with stony silence by official Turkey. Even his willingness (since late 1996) to accept autonomy or a federated solution rather than full independence did not draw a positive response. Turning its back on Communist ideology and symbols and adopting a social-democratic rhetoric helped the PKK in its public relations in Europe but not in eastern Turkey, where it has remained at loggerheads with the Turkish military.

Frustrated by Ankara's refusal to respond to their effort to move their struggle to the political arena, the Kurds transferred a major part of their activities to Europe. In 1995 they established a Kurdish parliament in exile. Some of its delegates were members of the Halkin Demokrasi Partici, the Kurdish political party in Turkey—which, as noted earlier, failed to achieve parliamentary representation in the 1995 elections.

With around a million Kurds living in Europe (more than half in Germany), PKK activities flourished there. The PKK also enjoyed the support of a number of European social-democratic parties, many holding the reins of government in their respective countries and well represented at the European Parliament. They were sympathetic to both the cause of Kurdish human rights and the PKK's willingness to conduct a political rather than a guerrilla struggle.

Kurdish attacks on Turkish targets continued in 1997 and 1998, and Turkey retaliated repeatedly inside Iraqi territory. Finally, in late 1998 the Turks decided to apply even greater pressure on Syria, as there was no more doubt that it was the main base of support for the Kurds' activities. Demanding that Syria cease its assistance to the PKK, declare it a terrorist organization, and force Öcalan out, the Turks concentrated troops on their border with Syria and threatened military moves. Tensions reached their peak in October 1998, to the extent that the Israelis saw fit to announce that they had nothing to do with the conflict and even reduced their troop movements in the Golan Heights to reassure the Syrians of their neutrality on the issue. Egypt, later joined by Iran, hastened to offer its services as mediator.

On 20 October 1998 an agreement was signed between Turkey and Syria in the Turkish town of Adana, which amounted in effect to a total Syrian capitulation. Syria admitted (after years of denials) that Öcalan was living in its territory, declared that it now regarded the PKK as a terrorist organization, and expressed its willingness to set up a joint security apparatus to check PKK activities in Syria and Lebanon. The agreement was signed by Chief of Political Security General 'Adnan Badir al-Hassan for Syria and Deputy Undersecretary of the Foreign Office Uğur Ziyal for Turkey.

As an immediate result, and for Turkey a very satisfactory one, Abdullah Öcalan was ordered to leave Syria at once. In early December it became known that he had arrived in Moscow and asked Russia for political asylum. Turkey was furious. Prime Minister Yılmaz rushed to meet Foreign Minister Igor Ivanov and demand that Öcalan be extradited to Turkey. Russia's parliament approved Öcalan's request for asylum, but the government preferred to deny it, and Öcalan found himself on a plane to Rome, where he was detained by the authorities.

Now Turkey exerted heavy pressure on the Italians, who claimed they were legally bound not to extradite Öcalan to Turkey, where he could expect the death penalty. The Turkish public was outraged; a boycott of Italian goods emerged spontaneously. Italy considered extraditing Öcalan to Germany, where he was also wanted for questioning, but the Germans (perhaps mindful of the fact that more than half a million Kurds were living in their land) prudently desisted from making a formal extradition request.

Thus Öcalan became a hot potato: no one in Europe wanted him, and no one was willing to hand him over to the Turks. The Italians soon found a way to get him out of their country. After traveling from one European airport to the next, in early February 1999 he surfaced at the Athens airport, asking various countries for asylum, including Greece. All requests were denied.

In the media, Libya and South Africa were mentioned as some of the places where Öcalan might go, but his movements were shrouded in secrecy. As determined journalists doubled their efforts to trace him, the world was suddenly informed that he had been abducted by Turkish security from his hideout in Kenya and brought to Turkey. The Turks were elated; Kurds all over the world were stunned.

Initial information given to the public tried to portray the Öcalan arrest as an operation conducted exclusively by the Turks. The international media, however, mentioned Israeli and U.S. assistance; subsequently, Turkish leaders, including Prime Minister Bülent Ecevit, did not deny that a foreign intelligence service was involved in locating Öcalan. In fact, former prime minister Yılmaz had already admitted that a foreign service had been helpful in informing the Turks of Öcalan's arrival in Moscow.

In Europe the Kurds felt betrayed—especially by Greece. Indeed, it seems the Greeks could have handled the entire affair much better than they did. While Öcalan was still "in transit" at the Athens airport, it was decided to move him to northern Greece and then put him on a flight to Nairobi. For two weeks Öcalan was sequestered at the residence of the Greek ambassador there while the Greeks were trying to find a refuge for him anywhere but in their own land. Yet Kurdish immigrants in Europe accused Greece of turning Öcalan in to Turkey, and a series of protest demonstrations—some of them violent—took place outside Greek embassies all over Europe. Israel was also held for blame. A particularly serious incident took place at the Israeli consulate in Berlin. Several Kurdish demonstrators managed to enter the building, where three of them were fatally shot by Israeli security guards and dozens were injured.

The bloodshed only intensified the media coverage of the Öcalan affair. He was awaiting trial in jail at Imrali Island in the Straits of Bosporus. There were also a few bomb attacks in Turkey proper, for which the PKK took credit.

After a few weeks things calmed down—but not for long. When Öcalan's trial began on 31 May 1999, the Kurdish issue flared up again. Initially, the three-man tribunal included a military officer. This raised a storm of protest in Europe, and consequently (as the trial was already in progress) Turkey decided to change the bench. It took a special parliament resolution, but the military man was replaced. Nevertheless, the revamped court sentenced Öcalan to death, and a wave of protests ensued once again. But no date was set for the execution, and it was impossible to tell at the time of writing when or if such a date would be set.

Turkey's treatment of its Kurdish population was exposed for the entire world to see. The long-standing policy of assimilation has been effective for about two-thirds of Turkey's Kurds, who now regard themselves as Turks for all intents and purposes. This, however, constitutes only a partial success as long as one-third of this ethnic minority still clamors for cultural and even national rights and gives passive or active support to the struggle of the Kurdish underground. Worldwide, too, numerous Kurdish exiles support the PKK, many of them openly. They are determined to make Turkey change its policies—to allow the teaching of the Kurdish language in schools and the giving of Kurdish names to children and to permit the publication of Kurdish-language newspapers. Some of the exiles continue to demand self-determination for the Kurds.

Official policy, which is shared by all major political parties, still rests on the twin pillars of assimilation on the one hand and a military solution on the other. It denies all allegations of human rights violations, claiming the Kurds are Turks in the eyes of the law. This is unacceptable to liberal public opinion in the West and elsewhere, and no solution has emerged.

The 1999 General Elections

In late 1998 several more corruption scandals shook Turkey, on top of the seemingly unending Susurluk affair discussed earlier. Eventually, the Yılmaz government was unable to withstand this relentless pressure. The direct reason for its downfall in November 1998 was the Korkmaz Yiğit affair.

The charges leveled at Yılmaz were rather serious. Businessman Korkmaz Yiğit, a rising star in the Turkish economy, was a personal friend and supporter of Yılmaz. It was alleged that the prime minister (and also the minister of state for economic affairs) had encouraged Yiğit to acquire the bank Türkbank, intended for privatization, for $600 million. Yiğit (who also acquired several newspapers and a television station) was a close associate of a Turkish organized crime boss, Alaattin Çakıcı, then in police custody in France. This immediately raised suspicions that Yiğit was in effect a "straw man" in a deal designed to sell a major Turkish bank to the mob—with the blessing of the prime minister, no less.

Thus overnight Yılmaz lost the enormous popularity he had acquired so recently when he brought about the deportation of Öcalan from Syria by applying unprecedented pressure on the Syrians, as noted earlier. Turkish public opinion was fed up with the numerous corruption scandals that had shaken the country in recent years, many of which involved former prime minister Çiller. Reaction was fierce, parliament joined the protest, and the government collapsed. Bülent Ecevit was asked to set up a caretaker government until the general elections, scheduled for 18 April 1999.

The main beneficiary (in terms of personal prestige) of the Öcalan deportation was Ecevit. It was during his brief term as caretaker that Turkey managed to locate the rebel leader in his hideout in Kenya. When he was brought to jail in Turkey, a wave of national pride swept the country. Thus the ideologically left-wing Ecevit won another nationalist laurel to add to his previous achievement, the invasion of northern Cyprus in 1974.

Nevertheless, the results of the April elections provided yet another stunning surprise. If Turkey had been shaken in December 1995 by the victory of the Islamic Welfare Party, now it was astonished by the Milliyetci Hareket Partisi (MHP, Nationalist Movement Party), which in 1995 had been unable to cross the 10 percent voting threshold and in 1997 lost its veteran leader, ex-colonel Alparslan Türkeş. Now the MHP won 19 percent of the total vote and 129 parliamentary seats, coming in second after Ecevit and the Demokratik Sol Parti (Democratic Left Party), which won 22 percent of the vote and 136 seats.[3] Ecevit's comeback, although impressive, was rather expected. The main surprise in 1999 was the great achievement of the extreme right-wing MHP, which this time had landed the protest vote.

The former repository of disenchanted votes, the Islamic Fazilet Partisi, lost about a quarter of its representation, which left it with 110 seats in parliament. The two central parties most closely associated with corruption scandals, Yılmaz's and Çiller's, won only 83 and 82 seats, respectively, far behind the leading trio. Another remarkable result was the failure of the Cumhuriyet Halk Partisi (Republican People's Party), led by Deniz Baykal, to pass the threshold—the first time in modern Turkish history the party founded by Mustafa Kemal Atatürk remained out of parliament following general elections.

This was a major upset, considered against the history of Turkish politics. In the past, extreme political parties from both left and right had failed repeatedly in their efforts to win parliamentary representation because of the steep threshold and their endemic divisions. Now the pan-Turanist nationalists, whose main quarrel in the past had been with Atatürk's notion that Turkish nationalism should be confined to Asia Minor, won the hearts and minds of many Turks. During the 1970s the MHP had been regarded as essentially a fascist party. Its militia, the Gray Wolves, spread terror in Turkish cities and towns. Some of its founding members had been barred from politics as early as 1953 because of their opposition to Kemalism and their fundamentalist tendencies. Their leader, Türkeş, who won the endearing appellate "the Führer," was removed from the junta that had led the 1970 coup because of his "authoritarian proclivities." Also, the MHP tacitly and not so tacitly supported the right-wing terrorism that was one of the main causes of the 1980 coup. And now this party occupied center stage.

The MHP won its remarkable achievement only two years after its unchallenged leader, the charismatic Türkeş, passed away. In the mid-1970s he had managed to become part of the political system, serving briefly as deputy prime minister under Demirel. But since then he had led his party from one failure to the next. His relatively unknown successor, however, Devlet Bahçeli, a fifty-one-year-old academic, led the MHP to its greatest achievement to date.

What happened in Turkey to bring about such a transformation? Undoubtedly, the capture of Öcalan and the consequent rise of nationalistic sentiments were significant factors. But at least equally important was the decision by the European Union (EU) not to include Turkey among the eleven nations to be considered for full membership during the following years.

In 1998 the EU decided to consider the membership of five nations in 2002 and six more in 2005. Turkey was not included in either list. Moreover, one of the first five countries was Cyprus—that is to say, "Greek" Cyprus. It was a devastating blow, coming a few years after Turkey had taken economic, political, and strategic risks by fully supporting the coalition war against Iraq. At the time, as mentioned earlier, the desire to move closer to the West had been a major consideration, outweighing both economic and religious factors that should have caused Turkey to assume a neutral position in the West's conflict with Iraq. Now it seemed all those losses and sacrifices were in vain.

Thus in April 1999 Turkey had a brand-new parliament with an unpredictable balance of power among its factions. In April 1999 few international commentators could explain exactly who Devlet Bahçeli was, although they were able to say that he had become the second-most-important person in Turkish politics. And, of course, nobody could tell how stable the new system was going to be.

At the time of writing, it seems the surprising election result produced a stable left-right coalition, and stability was what Turkey needed most after all the upheavals. Ecevit's relaxed style of leadership and the moderate position adopted by the MHP have been chiefly responsible for the current period of political stability.

Earthquake

At 3:02 A.M. on 16 August 1999 Turkey was shaken up—literally. An earthquake measuring 7.4 on the Richter scale hit the densely populated northwest section of the country. Large parts of the towns of Izmit, Adapazarı, Gölcük, Yalova, and Çınarcık, on the eastern shores of the Sea of Marmara, were reduced to rubble within forty-five seconds. Parts of Istanbul, Bursa, and Bolu were also damaged. In all, about 25,000 people were dead or missing, tens of thousands were injured, and more than 100,000 buildings were destroyed. The scope of damage reminded many observers of World War II bombings; some even mentioned Hiroshima and Nagasaki.

Earthquakes are rather common in Turkey. During the twentieth century alone there were dozens of lethal ones, but none as severe as the August 1999 quake, one of the worst in history.[4]

Nations of the world responded quickly. Around sixty nations—including the United States, France, Italy, Israel, and Greece—sent rescue and medical teams, helped set up tent camps for homeless survivors, and did their best to alleviate suffering. The area hit by the quake, around 31,000 square kilometers, is home to about 20 million people. One-third of Turkey's national product was produced there. To mention just one example, the Izmit refinery, which had been scheduled to be privatized and sold for nearly $1 billion, was almost completely destroyed. Dozens of factories ceased operation, some permanently. Tourism was badly hit as well. As one observer put it, the quake hit the locomotive that was pulling the train of Turkey's economic development.

In earlier chapters I described how Turkey was caught unprepared by the 1973 energy crisis. The political and economic ramifications then included the downfall of two governments (Ecevit's in October 1979, Demirel's in a coup d'état in September 1980), severe shortages that brought the economy to a near standstill and caused much personal suffering, and a radical shift in foreign policy away from the West and toward the oil-producing nations and the East. Now in 1999, Turkey was caught unprepared once again.

The earthquake was followed by an avalanche of criticisms, allegations, recriminations, and rumors. On 20 August the *Turkish Daily News* reported that the governors of Izmit, Sakarya, and Yalova had been removed from office. The government denied the allegation. "The whole state machinery is in shambles," the newspaper wrote, "and the authorities cannot hide their incompetence." The next day *Cumhuriyet* voiced the same opinion: "State machinery too collapsed in this earthquake." Even the hitherto immune military was severely criticized for arriving late at the disaster areas and for lack of leadership and efficiency in its rescue operations. Newspapers spoke of the "paralysis that seized the top echelons" and of the "disappearance" of top generals during rescue operations.

President Demirel and Prime Minister Ecevit—the same leaders who twenty years earlier had tried to drag Turkey out of the energy crisis—made a supreme effort to reassure the public that everything possible was being done. Demirel emphasized that the disaster was literally a force majeure, and Ecevit argued that any government would have had a difficult time coping with a disaster of that magnitude. But the media, domestic as well as international, was relentless. The selective collapse of structures followed the lines dividing rich and poor and corruption and honesty. Many charges were leveled at construction contractors and against public officials who had let them get away with violating building codes. Corruption was already a raw nerve—two former prime ministers had been forced to vacate their offices because of corruption allegations—and this natural disaster proved that official crookedness sometimes costs lives.

Criticism was leveled from the outside as well. The *New York Times* wrote in an editorial on 22 August 1999: "It is not a question of stopping earthquakes or preventing death altogether. It is a question of preventing needless death, of keeping a city from killing itself when an earthquake strikes. Nature may be insuperable in some forms, but human neglect is not." The *Economist* (28 August) accused the Turkish government of failing to prepare rescue squads and of incompetence in mobilizing the military for rescue operations. "Left to their own devices," the newspaper wrote, "Turks stepped in and found . . . that cooperation could fill the gap left by the state. This . . . will undoubtedly give a boost to the confidence of ordinary Turks to stand up for the first time and demand . . . a better deal from their rulers."

There was criticism even within the government's ranks. Minister of Tourism Ergün Mumcu said "the civil defense system never collapsed—it had never existed in the first place."[5] Minister of Health Osman Durmuş chose to attack overseas aid—"The Turkish people would have felt more comfortable with doctors who share their culture"—for which the *Radical* called him "a racist ignoramus" (23 August).

Islamic organizations tried to use the mosques (not neglecting to mention that they were relatively unharmed by the quake) for relief operations. Leaders of the opposition Fazilet Partisi described the disaster as heavenly punishment for the closure of religious schools and sales of alcoholic beverages and demanded an inquiry into the Turkish Red Crescent's dereliction in the performance of its duties. The party paper *Zaman* wrote that Turkey was a country "after an earthquake and before a political quake." The government and military were so concerned that Islamic circles would use the quake for political advantage that they curtailed the operations of Islamic relief and welfare organizations. Victims paid the price.

The government, however, survived this crisis, in part because international aid was fast and massive. The world had been less than generous with Turkey during the 1990s. European assistance was almost stopped because of allegations of human rights violations, and U.S. aid—around $750 million at the start of the decade—was smaller by a factor of ten toward its end. Yet the quake proved that Western nations consider Turkey too important to be allowed to regress into economic backwardness, with all its ramifications. What had happened to Iran and Iraq was too relevant for Turkey to be ignored.

Besides, the Turkish economy was much stronger in 1999 than it had been in the 1970s. It took less than a year for Turkey to recover economically from the impact of the quake and thus to prove the stability of both its political system, its industrial infrastructure, and its international stature.

Notes

1. Bruce Maddy-Weitzman, *Middle East Contemporary Survey, 1996,* Vol. 20 (Boulder: Westview, 1988), p. 710.

2. Ibid., p. 340.

3. Election statistics as reported by the Turkish press and the *Turkish Official Gazette (Bulletin)*.

4. Yet another lethal earthquake hit the same region on 12 November 1999. About 800 people died.

5. *Ha'aretz,* 24 August 1999.

Acronyms

ANAP	Anavatan Partisi (Motherland Party)
AP	Adalet Partisi (Justice Party)
bbd	barrel per day
CHP	Cumhuriyet Halk Partisi (Republican People's Party)
CIF	cost, insurance, and freight
D-G	director-general
DP	Demokratik Parti (Democracy Party)
DSP	Demokratik Sol Parti (Democratic Left Party)
DYP	Doğru Yol Partisi (True Path Party)
EEC	European Economic Community
EU	European Union
FP	Fazilet Partisi (Virtue Party)
FY	fiscal year
GNP	gross national product
Hadep	Halkin Demokrasi Partici (People's Democracy Party)
IEA	International Energy Agency
ISA	Israel State Archives
km	kilometer
kWh	kilowatt-hour
mbd	million barrels per day
MEED	*Middle East Economic Digest*
MHP	Milliyetci Hareket Partisi (Nationalist Movement Party)
MSP	Milli Selamet Partisi (National Salvation Party)
NATO	North Atlantic Treaty Organization
NIEO	new international economic order
OAPEC	Organization of Arab Petroleum Exporting Countries

OECD   Organization for Economic Cooperation and Development
OPEC   Organization of Petroleum Exporting Countries
PKK   Parti-ye Kerkaran-î Kürdistan (Kurdistan Workers' Party)
PLO   Palestine Liberation Organization
RCD   Organization for Regional Cooperation and Development
RP   Refah Partisi (Welfare Party)
SUMED   Suez-Mediterranean pipeline
TPAO   Turkish National Petroleum Company

Selected Bibliography

Altuğ, Yılmaz. "Turkish Foreign Policy vis-à-vis Recent World Developments." *Annals of the Faculty of Law.* University of Istanbul, 1976, pp. 71–84.

Batu, Hamit. "New Developments in Turkish Foreign Policy." *Foreign Policy* (Ankara). 5, no. 4 (1976): 5–17.

Berberoglu, B. *Turkey in Crisis.* London: Zed Press, 1978.

Cigdem, Balim, ed. *Turkey: Social and Economic Challenges in the 1990's.* Leiden: Brill, 1995.

Cordesman, A. H., and A. R. Wagner. *The Iran-Iraq War.* Boulder: Westview, 1990.

Demirer, M. A. "Turkey's Economic Relations with the Middle East." Ankara: Heritage Foundation, October 1984.

Evriviads, Marios. "Hegemonic Alliances and Destablization in the Middle East—Israel, Turkey and the U.S.A." *Pacis* (Deffensor), no. 4, (January 2000).

Feroz, Ahmed. *The Making of Modern Turkey.* London: Routlegde, 1993.

Gruen, G. E. *Turkey, Israel, and the Palestinian Question, 1948–1960.* Ph.D. dissertation, Columbia University, 1970.

———. "Turkey's Relations with Israel and Its Arab Neighbours." *Middle East Review.* 17, no. 3 (spring 1985).

Gunter, Michael M. *The Kurds in Turkey, a Political Dilemma.* Boulder: Westview, 1990.

Hale, William M. *Turkish Politics and the Military.* London: Routledge, 1994.

Hershlag, Z. Y. *The Contemporary Turkish Economy.* London: Routledge, 1998.

Hiç, Mükerrem. "The Question of Foreign and Private Capital in Turkey" and "Economic Policies Pursued by Turkey." *Orient.* 21 (1980) and 23 (1982): 220–224, respectively.

Karaosmanoğlu, Ali. "Islam and Foreign Policy: A Turkish Perspective." *Foreign Policy* (Ankara). 12, no. 1–2 (June 1985): 64–78.

———. "Turkey's Security and the Middle East." *Foreign Affairs*. 62, no. 1 (1983): 157–175.

Karpat, Kemal. *Turkish Foreign Policy in Transition*. Leiden: E. J. Brill, 1975.

Kemal, Ayhan. "Turkey's Relations with the Arab World." *Foreign Policy* (Ankara). 4, no. 4 (1974): 5–20.

Kramer, H., and F. Muller. "Relations with Turkey and the Caspian Basin Countries." In *Allied Divided*, R. Blackwill and M. Sturmer, eds. Cambridge: MIT Press, 1977.

Kuniholm, Bruce R. "Turkey and the West." *Foreign Affairs*. 70, no. 2 (spring 1991).

Kürkçüoğlu, Omer. "Turkey's Attitude Toward the Middle East Conflict." *Foreign Policy* (Ankara). 5, no. 4 (1975): 23–33.

Landau, J. *Radical Politics in Modern Turkey*. Leiden: E. J. Brill, 1974.

Larrabee, Stephen. "U.S. and European Policy Toward Turkey and the Caspian Basin." In *Allied Divided*, R. Blackwill and M. Sturmer, eds. Cambridge: MIT Press, 1977.

Lewis, Bernard. *The Emergence of Modern Turkey*. Oxford: Oxford University Press, 1968.

———. "Turkey Turns Away." *New Republic*. (spring 1978).

Liel, Alon. "Turkish Water: Bringing Peace to the M.E." *Insight Turkey*. 2, no. 2 (April-June 2000).

Mango, Andrew. *Turkey: The Challenge of a New Role*. Westport: Praeger, 1994.

Menashri, David, ed. *Central Asia and the Middle East*. London: Frank Cass, 1998.

Mumcu, Uğur. *Rabita*. Ankara: Tekin Yayinevi, 1987.

Nahamani, A. *Israel, Turkey, and Greece*. London: Frank Cass, 1987.

Rand, Christopher. *Making Democracy Safe for Oil*. Boston: Little, Brown, 1975.

Robins, Philip. *Turkey and the Middle East*. New York: Council of Foreign Relations Press, 1991.

Robinson, D. *The First Turkish Republic*. Cambridge: Harvard University Press, 1966.

Rustow, Dankwart. "Turkey's Travails." *Foreign Affairs*. 58, no. 1 (fall 1979): 87.

Shmuelevitz, Aryeh. *Turkey's Experiment in Islamic Government, 1996–1997*. Tel Aviv: Moshe Dayan Center, Tel Aviv University, May 1999.

Tachau, Frank. "Turkish Foreign Policy Between East and West." *Middle East Review*. 17, no. 3 (spring 1985).

Wieker,W. F. "Turkey, the Middle East and Islam." *Middle East Review*. 17, no. 3 (spring 1985).

Index

'Abd-al-Nasser, Gamal, 191, 199, 205
Abu Khail, Sheikh, 113
Abu Sharif, Sa'ad al-Din, 70
Adalet Partisi (AP; Justice Party), 11,
    12, 42, 46, 50, 72, 76, 131, 135–137,
    141, 179, 182, 183, 185, 200, 209
Ağar, Mehmet, 227
Akbulut, Yıldırım, 162
al-'Ali Ahmad, 146
Allon, Moshe, 203, 204
Alpkaya, Emin, 176
Altuğ, Yılmaz, 134, 152
Anavatan Partisi (ANAP; Motherland
    Party), 13, 14, 148, 164, 219–221,
    224, 227
Ankara, Bayram, 129
'Arif, 'Abd-al-Salaam, 4
Asiltürk, Oğuzhan, 200
al-Assad, Hafiz, 19, 146, 194, 198
Atatürk, Mustafa Kemal, 1, 7, 8, 10, 15,
    61, 173, 145, 148, 183, 219, 221,
    222, 229, 230, 239
Ates, Atilla, 198
Aydınlık, 81
'Aziz, Tarik, 164

Baghdad Pact, 66, 165, 190, 194,
    203–205
Baghri, Muhammad Reza, 175
Bahçeli, Devlet, 240
Baker, James, 163
Bani Sadr, Abu al-Hassan, 169, 171

Barak, Ehud, 77, 215
Barazani, Mas'ud, 234
Batu, Hamit, 142
Baykal, Deniz, 43, 46, 50, 72, 81, 85,
    130, 131, 176, 219, 224, 239
Bayülken, Haluk, 68, 106, 140, 168,
    194
al-Baz, Usama, 214
Bektaşı (order), 230
Ben-Gurion, David, 202, 205, 206, 215
Ben-Zvi, Yitzhak, 202
Bingöl, Serbülent, 48
Bir, Çevik, 213
Bourgiba, Habib, 4
Bozer, Ali, 162
Bucak, Sedat, 227
Bush, George, 163
Bush, George W., 151

Çağlayangil, İhsan Sabri, 57, 58, 64, 65,
    95, 136, 141, 142, 178, 182, 200,
    201, 206
Çakıcı, Alaattin, 238
Carter, James, 65
Çatlı, Abdullah, 227
Cem, Ismail, 187, 198
CENTO, 186
Çetin, Hikmet, 81, 83, 136
Ceyhun, Ekrem, 170
Çiller, Tansu, 212, 219–222, 225, 227,
    233, 238, 239
Clinton, Bill, 174

Cohen, Ya'acov, 115
Cumalıoğlu, Fehmi, 200
Cumhuriyet, 22, 47, 50, 131, 167, 171, 242
Cumhuriyet Halk Partisi (CHP; Republican People's Party), 10, 11, 46, 47, 50, 56, 57, 76, 132, 134, 136–138, 141, 177, 182, 184, 209, 219, 224, 229, 239

Demirel, Süleyman, 4, 11, 12, 40, 45, 50, 66, 72, 73, 76, 79, 81, 87, 97, 108, 114, 116, 131, 132, 135–138, 141, 158, 164, 168, 170, 179, 181, 183, 184, 198, 208, 209, 212, 221, 225, 228, 234, 240–242
Demokratik Parti (DP; Democracy Party), 10, 11, 130
Demokratik Sol Parti (DSP; Democratic Left Party), 219, 224, 227, 239
Denizcier, İbrahim, 210
Denktaş, Rauf, 141
Dıblan, Sezai, 151
Dincerler, Vehbi, 14
Doğan, Lütfi, 59, 60, 95, 112, 182
Doğru Yol Partisi (DYP; True Path Party), 219–222, 227
Durmuş, Osman, 242

Ecevit, Bülent, 11, 12, 42, 46, 47, 50, 54, 55, 63 72, 73, 76, 77, 79–82, 84–87, 96, 107, 108, 129, 131–137, 141, 145, 149, 151, 153, 158, 170, 175, 176, 181–185, 209, 219, 224, 227, 237–242
Elekdağ, Şükrü, 141
Emre, Süleyman Arif, 11
Erbakan, Necmettin, 11–13, 46, 48–50, 60–64, 69, 95, 112, 130–132, 140, 141, 143, 148, 152, 153, 175, 176, 178, 181–183, 197, 200, 207–209, 220–230
Erdoğan, Recep Tayyip, 228, 230
Ergenekon, Yılmaz, 50
Ergil, Doğu, 149, 150
Erkaya, Güven, 223
Erkman, Hayrettin, 12, 114, 115, 169, 170, 208, 209, 211

Esenbel, Melih, 167
European Union (EU), 7, 17, 20, 160, 161, 207, 220, 224, 229, 240
Evren, Kenan 9, 12, 108, 145, 147, 148, 150, 209

Fahd ibn 'Abd al-'Aziz al-Sa'ud, 65
Fazilet Partisi (FP; Virtue Party), 175, 228–231, 239, 243
Feisal II (king), 205
Feisal ibn 'Abd al-'Aziz al-Sa'ud, 4, 59, 62, 137, 140
Fisher, Morris, 203
Friedman, Thomas, 225

Ghotbazadeh, Sada, 170, 171
Giritlioğlu, Esat, 45
Gökmen, Oğuz, 54
Gönensay, Emre, 214
Gümrükçüoğlu, İbrahim, 225
Günaydin, 54, 178
Güneş, Turan, 47, 62, 63, 131, 136, 140, 152, 207
Gürtuna, Ümit, 167
Güvendiren, Ekrem Esat, 211

Haile Selassie I, 206
Halkin Demokrasi Partici (Hadep; People's Democracy Party), 219, 235
Hamadani, 'Adnan Hussein, 57, 84, 85
al-Hassan, 'Adnan Badir, 236
Hassan Muhammad ibn Yusuf (Hassan II), 4, 138
Herzog, Hayim, 206
Herzog, Ya'acov, 205
Hiç, Mükerrem, 42, 188
Hürriyet, 54, 55, 62, 71, 74, 198
Hussein, Saddam, 17, 55, 69, 157–159, 182, 234
Hussein ibn Talal, 4

İnan, Kamran, 22, 42, 50, 63, 65, 69, 70, 72, 141, 182, 200
İnönü, ğsmet 2, 3, 138
International Energy Agency (IEA), 38, 45–51, 110, 111, 116, 135
İpekçi, Abdi, 81
Irmak, Sadi, 47, 177
Isıl, Erhan, 47, 48, 49

Islamic Conference, 137, 139–145, 149,
172, 194–195, 207
Ivanov, Igor, 236

Jalud, 'Abd-al-Salaam, 74, 80, 177, 188
Johnson, Lyndon B., 3

Kadiri (order), 9
Kaharaman, Ismail, 222
Kandemir, Nüzhet, 22, 186
Karadayı, Ismail, 223
Karaosmanoğlu, Ali, 146
Kassim, 'Abd al-Karim, 205
Kavakçı, Merve, 231
Kayra, Chaıt, 166
Kazan, Şevket, 222
Keçeciler, Mehmet, 14, 148
Kemal, Sa'id, 201
Kemal, Yaşar, 223, 234
Khaddam, 'Abd-al-Halim, 194
Khail, Abu (Shaikh), 209
Khal'atbari, 'Abbas, 66, 167
Khomeini, Ruholla, 90, 146, 169,
171–173
Kılıç, Erhan, 230
Kılıç, Selahattin, 48, 57, 58, 59
Kırca, Coşkun, 215
Kirkuk-Dörtyol pipeline, 39, 40, 55, 56,
58, 59, 68, 70, 85, 93, 108, 158, 172
Kissinger, Henry, 46, 49
Kivukoğlu, Huseyin, 198
Kohen, Sami, 71
Köprülüler, Teoman, 101
Korutürk, Fahri S., 58, 66, 142, 167,
168, 201
Külahlı, Cemal, 101
Kurdistan Democratic Party (KDP), 234
Kurdistan Patriotic Union (KPU), 234
Kurt, Fahrettin, 9
Kutan, Recai, 228–230
Kutay, Kemal 230

Levy, David, 226
Lewis, Bernard, 7

Ma'aruf, Taha, 40
Mahous, Ibrahim, 194
Menderes, Adnan, 160, 191, 193,
203–206

Milli Selmet Partisi (MSP; National
Salvation Party), 11, 12, 13, 14, 46,
48, 49, 50, 60, 61, 132, 151, 182,
208, 209, 229
Milliyet, 60, 62, 70, 71, 81, 87, 92, 169,
170, 198
Milliyetci Hareket Partisi (MHP;
Nationalist Movement Party), 11,
219, 239, 240
Milo, Yehuda, 211
Mubarak, Husni, 201, 202
Mumcu, Ergün, 242
Mumcu, Uğur, 147, 148
Mussavi, Hussein, 148, 173

Nakşibendi (order), 9
Nasrallah, Hassan, 214
new international economic order
(NIEO), 131, 134, 136, 137, 151,
178
North Atlantic Treaty Organization
(NATO), 2–3, 47, 69, 134, 150,
161–163, 170, 178–180, 194, 199,
203, 204, 220, 225
Nurcu (order), 9

Öcalan, Abdullah, 15, 18, 19, 20, 195,
198, 201, 213, 215, 233–240
Ökçün, Gündüz, 80
Organization for Economic Cooperation
and Development (OECD), 46, 47,
48
Organization of Petroleum Exporting
Countries (OPEC), 5, 43, 46, 47, 49,
54–58, 70, 95, 96, 104–106, 118,
129, 131, 132, 136, 144, 176, 184
Özal, Korkut, 9, 147
Özal, Turgut, 9, 13, 17, 22, 23, 113,
122, 123, 137, 148, 149, 159–162,
164, 173, 195, 201, 210, 230
Özener, Barlas, ix, 214
Özgür İnsan, 63

Palestine Liberation Organization
(PLO), 74, 142, 179, 192, 195, 201,
211
Pamuk, Orhan, 223
Parti-ye Kerkaran-î Kürdistan (PKK;
Kurdistan Workers' Party), 15, 16,

17, 18, 19, 156, 164, 175, 176, 195,
   196, 198, 222, 223, 227, 233–238
"Peace Pipeline," 122
Pehlavi, Muhammad Reza (Shah), 66,
   67, 68, 90, 166, 168, 171, 176, 206
Pelletreau, Robert, 234
Peres, Shimon, 212
Perle, Richard, 225
Putin, Vladimir, 174

Qaddafi, Mu'ammar, 77, 146, 177–180,
   188, 226

Rabin, Yitzhak, 212
Rabitat al-'Alam al-Islami (World
   Islamic Union), 147, 148
Rafsanjani, 'Ali Akbar Hashemi, 187,
   225
Refah Partisi (RP; Welfare Party), 13,
   14, 148, 174, 219, 225, 228, 229,
   239
Riad, Mahmud, 201

Sa'adabad Treaty, 2
Sabah, 198
Sadat, Anwar, 199, 200
al-Sa'id Nuri, 203, 205
Sasson, Eliahu, 202, 204, 205
Sasson, Moshe, 205
Sener, Abdullatif, 222
Sezer, Necdet, 198
Sezgin, İsmet, 113
Sükan, Faruk, 130

Süleymancı (order), 9
Sunay, Cevdet, 4, 138

Talabani, Jalal, 234
Talu, Naim, 53, 59
Tercüman, 72, 73
al-Tohami, Hassan, 140
Toker, Metin, 62
Tubel, Selma, 71
Türkeş, Alparslan, 70, 200, 219, 239,
   240
Turkish Daily News, 241
Türkmen, İlter, 114, 210

Ulusu, Bülent, 108, 109, 144, 147
Ünsal, Vedat, 130

Vance, Cyrus, 169

al-Wakil, Ibrahim, 55
Weizman, Ezer, 212

Yamani, Ahman Zaki, 61, 62, 63, 65
Yankı, 60
Yiğit, Korkmaz, 238
Yılmaz, Mesut, 151, 198, 210, 214,
   219–221, 224, 227, 228, 230, 234,
   236–239
Yumurtalık, 69, 73

Zade, Ahmad Mossavi, 171
Zaman, 243
Ziyal, Uğur, 236

About the Book

At the turn of the twenty-first century, modern Turkey remains torn between the secular heritage of its founder, Kemal Ataturk, and the political and social trends that challenge that legacy. Alon Liel traces the development of Turkey's current political environment, investigating the collapse of the country's economy in the 1970s, its recovery in the 1980s, its relationship with its Middle Eastern neighbors, and the dramatic political events of the 1990s.

Alon Liel, currently serving as director-general of Israel's Ministry of Foreign Affairs, is on leave from the Department of International Relations, Hebrew University.